BestMasters

Springer awards „BestMasters" to the best master's theses which have been completed at renowned universities in Germany, Austria, and Switzerland.

The studies received highest marks and were recommended for publication by supervisors. They address current issues from various fields of research in natural sciences, psychology, technology, and economics.

The series addresses practitioners as well as scientists and, in particular, offers guidance for early stage researchers.

André Röhm

Dynamic Scenarios in Two-State Quantum Dot Lasers

Excited State Lasing, Ground State Quenching, and Dual-Mode Operation

Foreword by Prof. Dr. Kathy Lüdge
und Prof. Dr. Eckehard Schöll, PhD

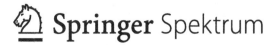 Springer Spektrum

André Röhm
Berlin, Germany

BestMasters
ISBN 978-3-658-09401-0   ISBN 978-3-658-09402-7 (eBook)
DOI 10.1007/978-3-658-09402-7

Library of Congress Control Number: 2015934951

Springer Spektrum
© Springer Fachmedien Wiesbaden 2015

Printed on acid-free paper

Springer Spektrum is a brand of Springer Fachmedien Wiesbaden
Springer Fachmedien Wiesbaden is part of Springer Science+Business Media
(www.springer.com)

Foreword

Semiconductor lasers are very small and cost efficient optical light sources that already entered into numerous parts of our daily lives. For the past 30 to 40 years, they have been used as experimental platforms for investigating new and interesting phenomena in physics and mathematics, and on the other hand as easy-to-handle test-beds for nonlinear dynamical phenomena occurring in a much wider context.

This thesis deals with the light emission characteristics of quantum-dot lasers, i.e., semiconductor lasers that contain pyramid-shaped nanostructures (quantum dots) coupled to a surrounding 2-dimensional layer of semiconductor material. Due to their multiple confined states these lasers can show multimode emission which is interesting for two reasons. At first it allows to study fundamental aspects of complex nonlinear dynamical system and second it may lead to technological applications in modern telecommunication where there is a need for multi-wavelength data transmission.

The microscopic modelling approach used within the thesis contains sophisticated microscopic modelling of the internal scattering processes and thus allows to quantitatively describe the light emission from the ground and the first excited state. As also observed in experiments, different operation modes are possible ranging from single mode operation to simultaneous two-state lasing or to a current depending ground state quenching. Especially the last phenomena, i.e., a shutdown of the ground state emission during an increase of the electric pump current, crucially depends on the internal scattering processes and is discussed in depth in the thesis. Supported by analytic approximations it is possible to predict the parameter regimes for ground state quenching and to identify the asymmetry in the electron and hole charge carrier populations as the driving force for the quenching.

The electrical modulation properties of a two-mode device can be significantly better than those of purely ground state emitting quantum-dot lasers. It is shown that an abrupt improvement is observed shortly behind the excited state emission threshold.

The thesis presents new results on the light emission characteristics of two-mode quantum dot lasers and suggests operation conditions for innovative and fast devices. This has considerable application potential, since quantum-dot lasers are very promising candidates for telecommunication applications and high-speed data transmission. Further the thesis gives new fundamental insights into the interplay between internal carrier scattering timescales and optical modulation properties by combining numerical solutions of nonlinear laser differential equations and analytical approximations methods.

Berlin, January 2015   Kathy Lüdge and Eckehard Schöll

Institute

The group of Prof. Eckehard Schöll at the Institut für Theoretische Physik of the Technische Universität Berlin has long-standing experience in nonlinear dynamics and control. In the center of the activities of the group have been theoretical investigations and computer simulations of nonlinear dynamical systems and complex networks with a particular emphasis on self-organized spatio-temporal pattern formation and its control by time-delayed feedback methods, and stochastic influences and noise. Recent research has focused on the deliberate control and selection of complex, chaotic, or noise-induced space-time patterns as well as synchronization and dynamics of complex delay-coupled networks, and chimera states. As state-of-the-art applications optoelectronic and neural systems are investigated, in particular coupled semiconductor lasers and optical amplifiers, nonlinear dynamics in semiconductor nanostructures, quantum dot lasers with optical feedback and injection, and neuronal network dynamics. The group is active in many national and international collaborations, and in particular is strongly involved in the Collaborative Research Center SFB 910 on Control of Self-Organizing Nonlinear Systems (Coordinator: Eckehard Schöll) and SFB 787 on Semiconductor Nanophotonics of the Deutsche Forschungsgemeinschaft (DFG).

Acknowledgement

First and foremost, I would like to thank the supervisors of my work, PD Kathy Lüdge and Prof. Eckehard Schöll, for valuable guidance and advice. Without their expertise and great lectures on nonlinear dynamics and on laser dynamics this work would not have been possible. Furthermore, I am especially thankful for the many fruitful discussions, suggestions and remarks by Benjamin Lingnau, who greatly helped me in getting through the trickier parts of numerical simulations, analytical approximations and the interpretation of results. I would also like to thank the other members of the nonlinear laser group, L. Jaurigue, C. Redlich, M. Wegert, F. Böhm, T. Kaul and R. Aust for constructive criticism and valuable advice. I am grateful for the seminars of our group AG Schöll, SFB787 and SFB910, which constantly exposed me to new ideas and often genuinely evoked my curiosity.

I would also like to gratefully acknowledge the fruitful cooperation with the work group of Prof. Woggon, especially Bastian Herzog and Yuecel Kaptan. Without their experimental expertise, the last part of this work would not have been possible. In that regard, fruitful input of H. Schmeckebier and N. Owschimikow is also acknowledged.

I am deeply grateful for the opportunity to publish my Master's Thesis in the form of this book and cannot thank Springer enough for their support and helpfulness.

Berlin, January 2015 André Röhm

Contents

List of Figures   **XIII**

List of Tables   **XV**

1. Introduction   **1**
   1.1. History   1
   1.2. Quantum Well and Quantum Dot Lasers   2

2. Theoretical Concepts of Lasers   **6**
   2.1. Basics of Laser Modelling   6
      2.1.1. Basic Concepts   6
      2.1.2. Cavity and Active Medium   8
      2.1.3. Laser Rate Equations 10
   2.2. Semiclassical Laser Theory 14
      2.2.1. Field Equations 14
      2.2.2. Matter Equations 17
      2.2.3. Modelling of Spontaneous Emission 20
   2.3. Model of a Quantum Dot Laser 22
      2.3.1. Dynamical Equations 22
      2.3.2. Scattering Rates 26

3. Modes of Operation of QD Lasers   **28**
   3.1. Single Colour Laser 28
   3.2. Two-State Lasing 30
      3.2.1. Experiments and Interpretation 30
      3.2.2. Numerical Simulations 31
   3.3. Ground State Quenching 33
      3.3.1. Experiments and Description 33
      3.3.2. Mechanisms of Quenching in the Literature 34

4. Understanding QD Laser Regimes of Operation   **37**
   4.1. Analytical Approximations 37
      4.1.1. Derivation 37
      4.1.2. Parameter Dependent Lasing Thresholds 38
   4.2. Numerical Simulation of GS Quenching 44
      4.2.1. Modeling Approaches and Light-Current Characteristics . . . 44
      4.2.2. Carrier Dynamics in GS Quenching 46
      4.2.3. Comparison with Analytical Approach 49
      4.2.4. Turn-On Dynamics 51
   4.3. Lasing Regimes In Parameter Space 52
      4.3.1. QD size and optical losses dependence 52
      4.3.2. Influence of Doping 54
      4.3.3. Temperature and ES gain dependence 56

5. Modulation Response   **60**
  5.1. Data Transmission with Semiconductor Lasers 60
  5.2. Modelling of Modulation 61
  5.3. Modulation Response Curves 62
  5.4. Cut-off-Frequencies and Two-State Lasing 66
  5.5. Ground State Modulation Enhancement 68
  5.6. Change of Cut-Off-Frequency with Carrier Loss Rates 71
  5.7. Outlook for Modulation Response 73

6. Pump-Probe Experiments   **76**
  6.1. Pump-Probe Setup 76
  6.2. Two-State Device Description 76
  6.3. First Experiment: Ground State Gain Recovery 77
  6.4. Second Experiment: Excited State Intensity Recovery 81

7. Summary and Outlook   **86**

Appendices   **89**

A. Scattering Rates   **89**
  A.1. Fully Non-Linear Rates 89
  A.2. Linearised Size-Dependent Scattering Rates 91

B. Deutsche Zusammenfassung und Ausblick   **93**

References   **95**

List of Figures

1.1. Energy band sketch of a p-n junction 2
1.2. Density of states versus energy for electronic states with different
      dimensions 3
1.3. Energy sketch of valence and conduction band of a quantum dot . . . 4
2.1. Sketch of the three fundamental single-photon interactions 6
2.2. Sketch of the pumping and energy level schemes of lasers 8
2.3. Sketch of a Fabry-Perot type laser 9
2.4. Steady states of the rate equation model 12
2.5. Turn-on time series for rate equations 13
2.6. Sketch of the confinement factor definition 16
2.7. Sketch of a QD emission spectrum with inhomogeneous broadening . 23
2.8. Sketch of the QD model used 25
3.1. Simulated light-current characteristic for single colour lasing 28
3.2. Simulated turn-on time series for single colour lasing 30
3.3. Simulated light-current characteristic for two-state lasing 32
3.4. Simulated light-current characteristic for two-state lasing with fast
      scattering 33
3.5. Simulated light-current characteristic for GS quenching 33
4.1. ES gain clamping and GS lasing regime in ρ_e^{ES}-ρ_h^{ES}-phase-space . . . 39
4.2. ρ_e^{ES}-ρ_h^{ES}-phase-space for lower energetic confinements 40
4.3. ρ_e^{ES}-ρ_h^{ES}-phase-space for stronger hole confinement 41
4.4. ρ_e^{ES}-ρ_h^{ES}-phase-space with energy asymmetry 41
4.5. ρ_e^{ES}-ρ_h^{ES}-phase-space for different temperatures 42
4.6. ρ_e^{ES}-ρ_h^{ES}-phase-space for different gains 42
4.7. ρ_e^{ES}-ρ_h^{ES}-phase-space for different ES gains 43
4.8. Simulated GS quenching by homogeneous broadening 45
4.9. Simulated GS quenching by self heating 45
4.10. Simulated GS quenching by hole depletion 46
4.11. Electron and hole densities versus current for hole depletion 47
4.12. Numerical results in ρ_e^{ES}-ρ_h^{ES}-phase-space 49
4.13. Numerical results for different scattering times in ρ_e^{ES}-ρ_h^{ES}-phase-space 50
4.14. Turn-on curves for two-state lasing 51
4.15. Lasing regimes versus current, confinement and optical losses 53
4.16. Lasing regimes versus current, confinement and optical losses with
      n-doping 55
4.17. Lasing regimes versus current, confinement and optical losses with
      p-doping 55
4.18. Lasing regimes versus current, temperature and ES gain 57
4.19. Lasing regimes versus current, temperature and ES gain with p-doping 58
4.20. Lasing regimes versus current, temperature and ES gain with n-doping 58
5.1. Two-state laser time series with modulated current 61
5.2. Two-state laser time series with modulated current; no relaxation
      scattering 62

5.3. Normalised modulation response versus frequency f 63
5.4. Modulation response of w_e, ρ_e^{GS} and ρ_e^{ES} versus frequency 64
5.5. Modulation response of w_e, ρ_e^{GS} and ρ_E^{ES} versus frequency with fits . . 66
5.6. 3dB cut-off-frequencies for two-state lasing 67
5.7. Analytical prediction for 3dB cut-off-frequencies for two-state lasing . 68
5.8. Absolute modulation response versus frequency f 69
5.9. 3dB cut-off-frequencies for high optical losses 70
5.10. 3dB cut-off-frequencies without GS capture scattering 71
5.11. 3dB cut-off-frequencies without relaxation scattering 72
5.12. 3dB cut-off-frequencies for fast scattering rates 72
5.13. 3dB cut-off-frequencies for high nonradiative losses 73
5.14. 3dB cut-off-frequencies for GS quenching 74
6.1. Sketch of a pump-probe experiment 76
6.2. Simulated light-current characteristic of the pump-probe experiment
device 77
6.3. Time-resolved GS gain recovery 78
6.4. Simulated GS inversion after pump pulse 79
6.5. GS recovery time for single-colour pump-probe experiments 79
6.6. Simulated GS inversion recovery time scales τ versus injection current 80
6.7. Simulated reaction of ES intensity versus time 82
6.8. ES intensity reaction fits; minimum depths and time versus current . 83
6.9. Simulated relaxation oscillation reaction 85
6.10. ES intensity reaction for resonant and off-resonant excitation 85

List of Tables

1. Dynamical variables of the numerical model 22
2. Parameters for the QD model. 25
3. Parameters used in the calculations for the single colour laser. 29
4. Parameters used in the calculations unless noted otherwise. 32
5. Numerical models used to study GS quenching 35
6. Parameters used in the calculations of section 4.1 39
7. Parameters used in the calculations of the experiments 84
8. GS capture scattering parameters 89
9. ES capture scattering parameters 89
10. Relaxation scattering parameters 90

1. Introduction

1.1. History

The 20th century saw the appearance of many new man-made materials, of which many never existed on earth before either in quantity or quality. Apart from purified radioactive elements used in atomic weapons or nuclear fission reactors and the petroleum derived polymer chemistry, semiconductor material sciences are among the most defining technologies shaping the later half of the last century [ELI14]. They not only gave rise to solar cells, diodes, LEDs and super-sensitive photodetectors, but grew to become the backbone of modern information technology since the first fabrication of a transistor by Shockley, Bardeen and Brattain in 1948. Silicon based microchips and controllers are used in smartphones, TVs, cars, planes, and even satellites and are therefore virtually omnipresent in our every day lives. The theoretical description of semiconductors had to match the rapid advancement made possible by ever more sophisticated fabrication techniques and the history of solid states physics is intrinsically linked with the development of quantum mechanics.

The concept for a semiconductor laser was first published by Basov *et al.* [BAS61] in 1961, only one year after the first experimental realisation of a laser by T. H. Maiman at Bell Laboratories in 1960 [MAI60]. But as opposed to this optically driven ruby laser, Basov *et al.* described a scheme for an electrically driven laser by recombination of charge carriers injected across a *p-n* junction. After some initial success, when in 1962 three independent groups produced the first semiconductor lasers, progress was slow [CHO99]. This was due to the fact that existing semiconductor technology was based solely on silicon. Silicon, however, does not exhibit a direct bandgap and is therefore not suited for use in laser systems. On the other hand, compound semiconductors were less well understood and fabrication was hard, so that the first lasers were only operable at cryogenic temperatures and only for a short pulse [BIM12]. For a more detailed history of the diode laser see Ref. [ELI14] and references therein.

The principle of an electrically driven semiconductor laser is shown in Fig. 1.1 (from [CHO99]). Here, the lower edge of the conduction band and the upper edge of the valence band is sketched together with the equilibrium electron density (shaded areas). With no voltage applied (a) the Fermi-energy is constant throughout the device and there are no regions where electrons could relax into unoccupied valence-band states. However, when a bias is applied in forward direction (b), electron-hole recombination becomes possible in the plane of the *p-n* junction.

This basic design is still used today in all electrically driven semiconductor lasers, commercial or otherwise. Among which there now exists a wide variety of different kinds, e.g. quantum well (QW) or quantum dot (QD) lasers. The great advantages of semiconductor lasers are not only low threshold currents and continuous wave (cw) output, but especially their small size, high temperature stability, room temperature operation and relatively cheap fabrication, all of which paved their way towards the wide use they are seeing today [BIM12].

This work will focus on a specific subset of semiconductor lasers, namely quantum dot lasers with two simultaneous lasing emissions. The stability of these lasing

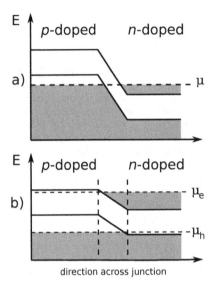

Figure 1.1: Energy band sketch of a p-n junction perpendicular to the junction plane (figure redrawn from *Semiconductor-Laser Fundamentals* by W. Chow and S. Koch, Springer (1999) [CHO99]). Without a voltage applied (a) the electrons (shaded areas) relax below the global Fermi-energy μ. For applied voltage in forward bias (b) electrons and holes can recombine at the p-n junction, enabling lasing.

states, originating from the ground state and excited state of the quantum dot, will be investigated temporally and parametrically. Especially the current induced appearance of excited state emission and subsequent quenching of ground state emission will be studied numerically and analytically. The suitability of these types of quantum dots for optical data transmission will also be touched upon, before the results of an experiment performed by the group of Prof. Woggon will be presented, numerically reproduced and interpreted.

1.2. Quantum Well and Quantum Dot Lasers

In 1963 Herbert Kroemer proposed [KRO63] to use a sandwich-like structure for the p-n junction. The charge carriers should be injected through an outer material layer with a high bandgap, while the active zone for the lasing should be fabricated from a smaller bandgap compound. The carrier density in this narrow region could then exceed the carrier densities of the injectors, which is impossible for a homogeneous structure. This enhances conversion of electrons into light and should be usable to reach higher quantum efficiencies and therefore lower threshold currents. There was, however, no sample fabricated at that time.

Independent of this, Charles Henry, who was working at Bell Laboratories, noticed that wave-guiding technology, which was used at that time to control the dominant direction of emitted laser light, could also be used to guide electron waves. The thickness, however, needed to be reduced below the de-Broglie wavelength of the electrons, so that the confinement became effective. A prototype was fashioned together with R. Dingle, exhibiting greatly reduced threshold currents. In 1976 a patent was filed, detailing the principle of quantum confinement for enhancing laser operations [DIN76].

When calculating the density of states of this structure, Henry obtained the remarkable result that it was not the known square-root dependence of three-dimensional structures, but a stepwise function due to discrete energy levels caused by the confinement in one direction (see Fig. 1.2). He concluded that this finite density of states, even for the lowest attainable energy, greatly enhanced laser operation. As only carriers of these lowest energies are participating in the electron-hole recombination, their increase in numbers leads to a higher optical gain and hence laser output for lower injection currents.

Obviously, charge carriers can be confined in more than one dimension, if different semiconductor heterostructures are grown. If the dimension is smaller than the de-Broglie wavelength in two directions, the resulting object is called a nanowire. When charge carriers are confined in all three dimensions, the structure was originally called quantum box and is nowadays called a quantum dot (QD). The resulting density of states is shown in Fig. 1.2. Each additional confinement increases the density of states at the lowest energy, and is followingly expected to increase laser performance.

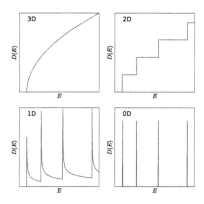

Figure 1.2: Density of states versus energy for electronic states with different dimensions. Lasing recombination usually involves the carriers of lowest energy, so the finite density of states even for the lowest occupiable state in the 2D case is favourable for laser operation. Going to even lower dimensions further strengthens this effect.

Quantum dots (QDs) even display a discrete spectrum of energy states. As this resulting spectrum, both for holes and electrons, is very similar to the discrete eigenstates of atoms, they are labelled in the same fashion: The lowest energy state is called ground state (GS), the next higher state the 'first excited state' (ES) and so forth. Generally, the number of confined states depends on the size of the quantum dot. Due to fabrication processes the confinement in reality is less than perfect. Therefore, the eigenstates of the quantum dots are usually modelled within a parabolic potential, however more advanced modelling approaches also exist [SCH07f]. Figure 1.3 shows a sketch of the energy bands and localised states for a quantum dot.

Confinement also gives rise to a zero-point energy, so that the energy separation

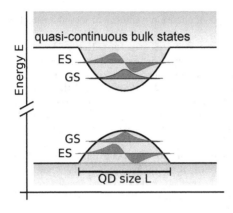

Figure 1.3: Energy sketch of valence and conduction band of a quantum dot for a parabolic potential. The QD is fabricated from a lower-bandgap material than the surrounding bulk material, the confinement in all three dimensions giving rise to discrete energy levels. Due to their similarity to atomic states, they are labelled ground state (GS), first excited state (ES) and so forth.

between electrons and holes for the GS of the nanostructure is not simply given by the band gap of the material. Hence, the resulting optical transition of frequency ω possesses the energy $\hbar\omega = E_{bandgap} + E_0$. As in the quantum mechanical example of a particle in a box, this zero-point energy E_0 is size-dependent. This effect is present in quantum wells, wires and dots and is often exploited for changing the wavelengths of emitted light by changing the confinement size L. Naturally, there are limits for tuning the wavelength like this, as structures can neither grow too big nor too small, and following we there is still a wide variety of material systems used today for obtaining lasers from infrared to ultraviolet.

Now, several theoretical groups predicted during the 1980s, that quantum dots would exceed even quantum wells in their performance [ARA82, ASA86], so that the attention was shifted towards fabricating these new types of nanostructures. QDs have yet to fulfil these promises, as it turned out that carrier injection into the quantum dot is a limiting factor for lasing operation and thus they remain a topic of major industrial and scientific interest even today.

This work will focus on typical self-assembled InGaAs quantum dots, though many of the findings presented here should also be valid for other material systems. Self-assembled InGaAs quantum dots are produced by Stranski-Krastanov growth either in molecular-beam epitaxy (MBE) or metal-organic vapour-phase epitaxy (MOVPE). In the critical step of this procedure, indium arsenide is grown on gallium arsenide in a thin layer, ranging from one to three monolayers. As Gallium arsenide and indium arsenide have a lattice mismatch of 7%, this leads to a highly stressed surface. During this stage and possibly during a short heating phase applied thereafter, the indium arsenide surface breaks open and reassembles itself into pyramidic structures. These pyramids are energetically more favourable, due to their tops being less stressed, for a certain temperature range during the growth process.

Afterwards, these QDs are overgrown with another layer of gallium arsenide, to form a dot-in-a-well structure [KOV03], which shifts the output wavelength to

1.3μm. As the creation process is stochastic, the sizes and shapes of the QDs are statistically distributed. Together with the inability to predict the exact spot where a QD will appear, this is arguably one of the the main disadvantages of contemporary QD technology. Their emission spectrum is broadened by their different sizes and compositions, and only a part of the QD ensemble is able to participate in the lasing process [BIM99, LUE11a, RAF11].

2. Theoretical Concepts of Lasers

2.1. Basics of Laser Modelling

Lasers are light sources with very narrow bandwidths, high output power and long coherence lengths [HAK86]. They are, in some regards, completely different from other light sources, that surround us every day. As opposed to the thermal radiation of light bulbs, stars and the sun, or the fluorescence used in neon tubes, the photons of a laser are mainly emitted through *stimulated emission*.

This section will explain the basic concepts of a laser and present a simple numerical model to show some general laser dynamics.

2.1.1. Basic Concepts

When in the early 20th century the particle-like nature of light was discovered and Niels Bohr formulated the famous Bohr model of the atom, two types of light-matter interactions were soon understood. Firstly, spontaneous emission is the stochastic decay of an excited electron, where a photon is emitted during the electrons transition from an upper state with energy E_2 to a lower energetic state of energy E_1. Secondly, an incoming photon of matching energy $\hbar\omega = E_2 - E_1$ can be absorbed while lifting the electron from the lower to the higher state.

However, Albert Einstein proposed the existence of an additional interaction, namely *stimulated emission*, in 1917 [EIN17]. There, an incoming photon of matching energy $\hbar\omega = E_2 - E_1$ is not absorbed, but encounters the electron in the upper state and stimulates the decay into the lower state. Hence, a second photon is emitted, which is identical in phase and direction to the first one. On a macroscopic scale the light intensity is coherently amplified by this process. Figure 2.1 shows a sketch of the three single-photon processes described above.

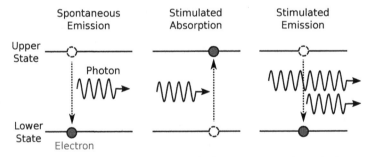

Figure 2.1: Sketch of the three fundamental single-photon interactions of a two-level system. For spontaneous emission (a), an electron (red) decays from the upper energetic state to the lower energetic state, while emitting a photon (blue). Conversely, through stimulated absorption (b) an electron is lifted into the upper state, while a photon is simultaneously consumed. Lastly, stimulated emission (c) is the coherent emission of a second photon, by an incoming photon that finds the electron in the upper state.

Through simple calculus [HAK86] Einstein could also show that the light-matter interaction coefficients, nowadays called in his honour Einstein-coefficients, had to be identical for stimulated emission and absorption. Stimulated emission is therefore often seen as the reverse process of (stimulated) absorption. Followingly, a net amplification of incoming light can only be achieved by stimulated emission, if more electrons are available for stimulated emission than for absorption, ergo if the population N of the upper state is higher than of the lower state, $N_2 > N_1$.

Yet, this state of 'population inversion' is never reached in thermal equilibrium. Mathematically, the Maxwell-Boltzmann-distribution only allows states with higher energies to be filled more, if the temperature is set to a negative value [HAK86]. Population inversion is therefore sometimes also referred to as 'negative temperature', albeit macroscopic systems never reach negative temperatures as a stable equilibrium state.

Hence, the system must be constantly driven out of thermal equilibrium to achieve 'population inversion'. This process called 'pumping' can be achieved through various ways and depends on the system being used. It can be optically, electrically or even chemically driven [HAK86]. Some pumping mechanisms will only provide sufficient inversion for a very short time leading to pulsed lasers, while others allow the emission of a continuous wave. Furthermore, even though only two energetic levels are participating in the optical transition, all real-world lasers do incorporate at least three different energy levels, often even four, and are congruently referred to as three-level lasers and four-level lasers [ERN10b]. A sketch of the pumping and lasing transitions of these systems is shown in Fig. 2.2.

The advantage of involving additional states for the electrons are easy to understand: While an optically driven two-level system can never reach population inversion through optical pumping, as the absorption and stimulated emission balance each other out, the three-level system avoids this by indirect excitation. The electrons are lifted from the lowest level of energy E_1 to a level above the upper level involved in the transition. In a suitable material, the electrons in this state of energy $E_3 > E_2$ then quickly decay into the upper state of energy E_2 of the lasing transition. An effective pumping without disturbing level 2 is therefore possible. Yet, to reach population inversion at least half of the population of level 1 would still have to be excited. This strict requirement is lifted for the four-level laser.

In the four-level system of energies $E_1 < E_2 < E_3 < E_4$, the electrons are lifted from the level 1 in to the level 4 through pumping, quickly decay into the metastable level 3, where they are used for stimulated emission, similar to the three-level system. However, in the optimal case the transition from energy level 2 to 1 is extremely fast, leaving the level 2 constantly almost empty and followingly keeping a population inversion between levels 2 and 3. Overall, the pumping requirements for lasing operations are greatly reduced and hence four-level systems are common [ERN10b, HAK85]. Figure 2.2 shows pumping and transition schemes for the three-level and four-level system.

Quantum dot lasers can be seen as a four-level system, with the conduction band and valence band acting as the highest and lowest level, while the quantum dot levels encompass the optically active lasing transition. However, the complex scattering dynamics involved lead to a variety of additional effects, e.g. two-state lasing or ground state quenching.

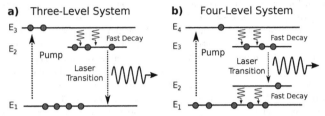

Figure 2.2: Sketch of the pumping and energy level scheme for the three-level laser (a) and four-level laser (b). Electrons (red) are raised by the pump to the highest energetic level, from which they quickly decay into the upper state of the lasing transition. This state is ideally metastable, so that electrons can accumulate there. For the three-level system (a), the lasing transition then links this metastable state E_2 to the ground level E_1. The four-level laser (b) possesses an additional level E_2, that acts as the lower state for the lasing transition. Because this level E_2 is short lived, it is almost always empty, leaving the E_3-E_2 transition easily inverted.

From an engineering perspective, the laser converts the energy injected into the system via pumping, e.g. the injection current in a semiconductor laser, into coherent light. One can therefore easily formulate conversion efficiencies by measuring the output versus input power. Electrically driven semiconductor lasers are among the most efficient lasing systems [CHO99].

Mathematically, the amplification of an incoming electro-magnetic wave is often measured as *gain g*. In a simple model the electric field amplitude E will increase exponentially over time with gain g:

$$\frac{d}{dt}E = gE. \tag{2.1}$$

Gain is therefore quantified in units of [1/s]. Microscopically, the gain of a medium is related to its population inversion:

$$g \sim N_2 - N_1, \tag{2.2}$$

where N_2 and N_1 are the populations of the upper and lower electronic level. When the lower level is more populated, the gain g becomes negative and instead of amplifying the incoming wave, the medium becomes absorbing. Gain g can therefore also be seen as an inverse absorption coefficient. The interplay of gain and light will be discussed in more detail in the following section.

2.1.2. Cavity and Active Medium

The two principal components of every laser are the *optical cavity* and the optically active *gain medium* [HAK85] contained inside. The cavity is a confined space in which certain standing electromagnetic waves can exist. These 'cavity modes' have a discrete set of eigenfrequencies and can be excited via injection of photons. Typically the edge of the cavity will be a mirror or another reflecting surface, so that photons of the cavity modes are reflected. The light then passes through the gain medium multiple times before being absorbed or escaping the cavity. The gain medium is a material, which amplifies light through the process of stimulated emission in the manner described in the previous section. When placed in a cavity, it will be exposed to its own amplified emission and create a coherent, intensity amplified standing wave. This is the origin of the name laser, an acronym for 'Light Amplification by Stimulated Emission of Radiation'.

However, the gain medium does not enhance all optical frequency equally, but possesses a gain profile. This gain profile usually corresponds to the spontaneous emission spectrum of the optical transition that is used for amplification, e.g. a Gaussian shape with its natural line width. For lasers there are usually many cavity modes lying within the peak of the gain spectrum, so that the laser, in principle, could operate on many different modes. Hence, further mode selections becomes necessary.

One easy way of mode selection is to use a Fabry-Perot resonator, as proposed by Schawlow and Townes [SCH58] in 1958. There, only modes along the principal axis of the resonator are enhanced. It consists of two parallel mirrors and significantly reduces the number of modes remaining inside the gain spectrum for optical amplification. A sketch of a Fabry-Perot type laser with all integral components can be seen in Fig. 2.3. Additionally, the coherent light has to be coupled out of the cavity for further use. This is achieved by using a high-reflectivity mirror on one side, and a low reflectivity mirror on the other. Light will then be mainly exiting through the low reflective end of the cavity.

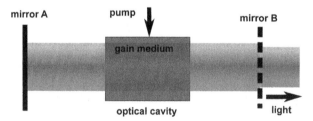

Figure 2.3: Sketch of a Fabry-Perot type laser. The cavity, also called optical resonator, consists of two planar parallel mirrors. The reflectivity of mirror B is lower than of mirror A, so that light mainly leaves the cavity on this side. Inside is the gain medium, which is kept in a state of inversion via pumping. Stimulated emission leads to the appearance of one stable coherent standing wave mode inside the resonator.

In this work the focus lies on the Fabry-Perot type of devices, and it will be assumed that only one lasing mode exists, corresponding to the mode with the highest optical amplification. Excluding the multi-mode dynamics not only decreases computational costs and complexity, but is also consistent with most fabricated QD laser structures, e.g. those with distributed Bragg-reflectors (DBRs). Usually, the end of the semiconductor sample is simply cleaved and the resulting air-bulk material interface is used as a mirror, resulting in a Farby-Perot type cavity. The alignment is automatically generated by splitting both ends among the same crystallographic plane.

2.1.3. Laser Rate Equations

There are many ways of mathematically describing a laser, suited for different material systems, time scales and types of lasers. However, some laser properties are universal and can hence be understood with even the simplest model approach.

Before the semiclassical laser-equations will be derived in the next section, a simple two-variable rate-equation model shall be heuristically motivated and studied here. The specific set of differential equations are taken from T. Erneux and P. Glorieux [ERN10b] and represent such a minimal laser model. They are given by:

$$\frac{d}{dt}I = ID - I$$
$$\frac{1}{\gamma}\frac{d}{dt}D = (A - D) - DI, \tag{2.3}$$

where I is the light intensity and D is the inversion of the gain medium. Both dynamic variables are normalised, to have as few parameters remaining as possible. As can be seen, the time evolution of the light intensity $\frac{d}{dt}I$ contains a normalised decay term $-I$, which models the loss of light due to absorption and transmission at the mirrors. Furthermore, the product term ID simulates the stimulated emission, which is stronger for more light and higher population inversion and hence linear in both I and D.

This increase in light intensity translates into a loss of inversion D, as stimulated emission uses up excited carriers. Therefore, $-ID$ enters the time evolution of D as a loss term. Additionally, the inversion is being externally driven towards a static value, prescribed by the pump parameter A. The specific nature of the pump is not further specified - it is simply assumed that through some mechanism the inversion of the gain medium can be excited. Lastly, γ is a parameter describing the time scale separation between carrier and light dynamics, usually in the range of 10^{-2} to 10^{-6} [ERN10b].

Now, it is easy to find the steady states of this simplified system by solving the equations:

$$ID - I = 0$$
$$(A - D) - DI = 0. \tag{2.4}$$

There are two sets of steady states fulfilling these conditions. The first is given by:

$$I^{off} = 0$$
$$D^{off} = A. \qquad (2.5)$$

With the light intensity I at zero and the inversion D at pump level, this represents an 'off'-state. No intensity is produced and carriers are dominated by the external pumping A. Conversely, the second steady state solution of Eq. (2.4) yields the 'on'-state:

$$I^{on} = A - 1$$
$$D^{on} = 1. \qquad (2.6)$$

Here, the light intensity I is proportional to the pump parameter A, caused by the conversion of injected energy into lasing light. Simultaneously the inversion is constant with $D^{on} = 1$. This effect is called *gain clamping* and is a result of the stimulated emission dominating the system. As can be seen from the differential equation for I, $D = 1$ is the transparent state of the system, where the stimulated emission and optical losses cancel each other out.

Now, a linear stability analysis of the system can be calculated. The Jacobian $\underline{\underline{J}}$ of Eq. (2.3) is given by:

$$\underline{\underline{J}} = \begin{bmatrix} D - 1 & I \\ \gamma D & \gamma(-1 - I). \end{bmatrix} \qquad (2.7)$$

With this Jacobian, the time evolution of small perturbations δI and δD around the steady states can be described:

$$\delta I = I - I^{on,off}$$
$$\delta D = D - D^{on,off}$$
$$\frac{d}{dt}\begin{pmatrix} \delta I \\ \delta D \end{pmatrix} = \underline{\underline{J}} \begin{pmatrix} \delta I \\ \delta D \end{pmatrix} + \mathcal{O}\left(\delta I^2, \delta D^2\right), \qquad (2.8)$$

where a vector notation was used for δD and δI. The linear differential Eq. (2.8) can be solved with a two-exponential ansatz:

$$\begin{pmatrix} \delta I \\ \delta D \end{pmatrix} = \begin{pmatrix} a_1 \\ a_2 \end{pmatrix} e^{\lambda_1 t} + \begin{pmatrix} b_1 \\ b_2 \end{pmatrix} e^{\lambda_2 t}, \qquad (2.9)$$

where a and b are coefficients for the initial value, and $\lambda_{1,2}$ are the eigenvalues of the Jacobian $\underline{\underline{J}}$. Now, when the steady state variables $I^{on,off}$ and $D^{on,off}$ are inserted, the eigenvalues of the resulting matrix can be easily calculated. For the off-state they are:

$$\lambda_1^{off} = A - 1$$
$$\lambda_2^{off} = -\gamma. \tag{2.10}$$

For values of $A < 1$, both eigenvalues are negative. Following Eq. (2.9) this means that all small perturbations δI and δD decay exponentially, so that the steady state is stable. Conversely, for values of $A > 1$, hence for stronger pumping, the *off*-state becomes unstable. The eigenvalues for the *on*-state are given by:

$$\lambda_{1,2}^{on} = -\gamma\frac{A}{2} \pm \sqrt{\gamma^2 A^2/4 - \gamma(A - 1)}, \tag{2.11}$$

for which the plus-combination changes sign. For $A < 1$ the *on*-state is unstable, while for $A > 1$ it is stable. Figure 2.4 shows the steady state solutions of Eq. (2.5) and Eq. (2.6). The *off*-state and *on*-state exchange stability in a transcritical bifurcation at $A = 1$.

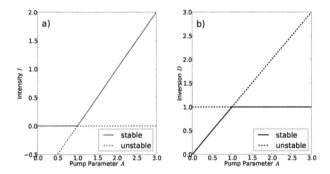

Figure 2.4: Steady states of the rate equation model, (a) intensity I and (b) inversion D against pump parameter A. The stable solutions (solid lines) switch in a transcritical bifurcation at $A = 1$. For $A < 1$ the laser is turned off ($I = 0$) and inversion increases linear with the pump A. The *on*-state is stable for $A = 1$, and $D = 1$ is *gain-clamped*.

As the lasing intensity (a) is zero before, and increases linearly afterwards, $A = 1$ is called the *lasing threshold* and is a typical feature of laser dynamics. On the lasing threshold the system undergoes a change of stability and the state of the system is qualitatively different afterwards. For this simple two-variable rate equation approach here the intensity is zero below threshold and this drastic transition is followingly quite obvious, but even in more complex systems with spontaneous emission included, a pump current corresponding to the lasing threshold can be identified [ERN10b]. It marks the transition towards stimulated emission and the onset of coherent light emission.

Figure 2.4 (b) also shows the *gain clamping* of the inversion and visualises the lasing threshold in terms of carrier dynamics. So while for $A < 1$ the inversion increases linearly with pump A, which is the result of more electrons getting excited through the pumping mechanism, this rise is suddenly stopped at the lasing threshold $A = 1$. As mentioned, for $D = 1$ the stimulated emission cancels out the decay of intensity I, so that the lasing intensity is stable. If the system were to reach a state of $D > 1$, this would result in an amplification of I through stimulated emission. However, an increased lasing intensity I also increases the losses that the stimulated emission term $-DI$ represents for the time evolution of D in Eq. (2.3). So while the lasing intensity goes up, inversion is consumed simultaneously. Followingly, there are no steady states with $D > 1$, as any excess inversion is always converted into increased light intensity.

Gain clamping above the lasing threshold is a key feature of all lasing systems. If enough carriers get excited, the stimulated emission will start to dominate the system. This always suppresses the participating inversion to a state of transparency, where stimulated emission and optical losses cancel each other out.

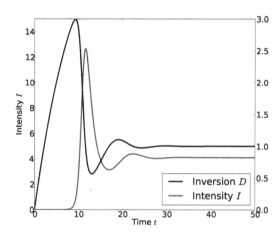

Figure 2.5: Turn-on time series for the set of rate equations obtained by numerical integration. For pump parameter $A = 5.1$ and $\gamma = 10^{-1}$ relaxation oscillations are clearly visible, caused by the periodic interaction of intensity I (red) and inversion D (blue).

Lastly, for $\gamma < 1$ the eigenvalues of Eq. (2.11) turn into a pair of complex conjugated eigenvalues. So the *on*-state is actually a stable focus. This gives rise to *relaxation oscillations*, which is the periodic exchange of energy between gain medium and light field during turn-on. Figure 2.5 plots the turn-on for the rate equation model above threshold ($A = 5.1$) for $\gamma = 10^{-1}$. Both the inversion D and intensity I clearly overshoot and then exhibit damped oscillations before converging towards their steady state values.

The frequency ω_{osc} and damping Γ of the relaxation oscillations is given by the real and imaginary part of the eigenvalue of Eq. (2.11):

$$\Gamma = \gamma \frac{A}{2}. \qquad (2.12)$$

The damping increases linearly with pump A, while the frequency can be expanded for small γ to [ERN10b, OTT14]:

$$\omega_{osc} = \sqrt{-\gamma^2 A^2/4 + \gamma(A - 1)},$$
$$\simeq \sqrt{\gamma(A - 1)} + \mathcal{O}(\gamma^{3/2}). \tag{2.13}$$

Hence, relaxation oscillations are slowest at the threshold $A = 1$, but also least damped. This result once again is also true for more complex laser systems[LUE11]. Semiconductor lasers also exhibit relaxation oscillations, which can be used for the generation of short pulses by gain-switching[SCH88j]. However, the relaxation oscillations of QD semiconductor lasers as studied in this work are often so strongly damped [ERN10b], that they are not even visible. This is important for a range of properties, e.g. their stability against perturbations and modulation dynamics.

Overall, such a simple rate-equation model is useful for outlining and visualising a variety of general laser properties. However, two-state lasing quantum dots are not reproducible, as one needs to include more carrier reservoirs and lasing fields. Furthermore, a theoretical description should be derived from first principles, to ensure that all important aspects are taken into account and experiments can be accurately modelled. Therefore, the next section will cover the semiclassical laser-equations.

2.2. Semiclassical Laser Theory

2.2.1. Field Equations

To accurately model a semiconductor laser, equations of motions for the electric field and the internal states of the gain medium must be derived from first principles. Because the number of photons is very large, it is sufficient [LIN11b, MEY91] to treat the field equations classically for most applications, while the gain medium is treated in the framework of quantum mechanics. This leads to the Maxwell-Bloch equations of semiclassical laser theory.

First, the field dynamics shall be derived. As a starting point, Maxwell's equation in matter are given by [HAK86]:

$$\operatorname{div} \boldsymbol{D} = \rho, \tag{2.14}$$
$$\operatorname{div} \boldsymbol{B} = 0, \tag{2.15}$$
$$\operatorname{rot} \boldsymbol{E} = -\dot{\boldsymbol{B}}, \tag{2.16}$$
$$\operatorname{rot} \boldsymbol{H} = \boldsymbol{j} + \dot{\boldsymbol{D}}. \tag{2.17}$$

Here $\boldsymbol{B}$ and $\boldsymbol{E}$ are the magnetic induction and electric field strength. ρ is the density of free electric charge carriers, $\boldsymbol{j}$ the corresponding current density. The dielectric displacement $\boldsymbol{D}$ is connected to $\boldsymbol{E}$ via

$$\boldsymbol{D} = \epsilon_0 \boldsymbol{E} + \boldsymbol{P}_{all}, \tag{2.18}$$

where ϵ_0 is the vacuum permittivity and $\boldsymbol{P_{all}}$ the polarization of the medium. The polarization will now be split into a resonant part $\boldsymbol{P_r}$ and an off-resonant background polarization $\boldsymbol{P_{bg}}$:

$$\boldsymbol{P_{all}} = \boldsymbol{P_r} + \boldsymbol{P_{bg}}. \tag{2.19}$$

While the resonant polarization $\boldsymbol{P_r}$ will need to be modelled microscopically, the off-resonant polarization $\boldsymbol{P_{bg}}$ is assumed to act linearly during laser operation:

$$\boldsymbol{P_{bg}} = \epsilon_0 \chi_{bg} \boldsymbol{E}, \tag{2.20}$$

and will be absorbed into $\epsilon_{bg} = 1 + \chi_{bg}$. For non-magnetic materials, the magnetizing field $\boldsymbol{H}$ is given by

$$\boldsymbol{B} = \mu_0 \boldsymbol{H}. \tag{2.21}$$

With no free charge carriers $\rho = 0$ and no free current $\boldsymbol{j} = 0$ the Equation for the electric field can be derived as:

$$\Delta \boldsymbol{E} - \frac{n^2}{c^2} \ddot{\boldsymbol{E}} = \mu_0 \ddot{\boldsymbol{P}}, \tag{2.22}$$

where the subscript of $\boldsymbol{P_r}$ has been suppressed and the relation $n^2 c^{-2} = (\epsilon_0 \epsilon_{bg} \mu_0)^{1/2}$ was used. Here, c denotes the vacuum speed of light, while $n = \sqrt{\epsilon_{bg}}$ is the refractive index of the medium and $\Delta = \partial_x^2 + \partial_y^2 + \partial_z^2$ denotes the Laplace-operator.

To further simplify, it will now be assumed that the electric field $\boldsymbol{E}$ can be approximated as a plane wave in z-direction with frequency ω, wave number k and envelope amplitude function $E(t)$. This is justified, as the laser possesses a dominant direction and inside the resonator standing waves are formed. A corresponding approach is taken for the polarization $\boldsymbol{P}$:

$$\boldsymbol{E}(z,t) = \hat{e}_x E(t) \exp\left[i\left(kz - \omega t\right)\right] \tag{2.23}$$

$$\boldsymbol{P}(z,t) = \hat{e}_x P(t) \exp\left[i\left(kz - \omega t\right)\right] \tag{2.24}$$

Here, $\hat{e}_x$ denotes a constant unit vector in the direction of polarization and $E(t)$ is the time-dependent electric field amplitude. After inserting Eq. (2.23) and Eq. (2.24) into Eq. (2.22) the following equation for E and P is obtained:

$$k^2 E - \frac{n^2}{c^2}(\ddot{E} - 2i\omega \dot{E} - \omega^2 E) = -\mu_0\left(\ddot{P} - 2i\omega \dot{P} - \omega^2 P\right). \tag{2.25}$$

Now the dispersion relation in matter is used:

$$k^2 = \frac{\omega^2 n^2}{c^2}, \tag{2.26}$$

which leads to the following equation of motion for the envelope function of polarization P and electric field E:

$$\ddot{E}(t) - 2i\omega\dot{E} = -\frac{1}{\epsilon_0\epsilon_{bg}}\left(\ddot{P}(t) - 2\omega\dot{P} - \omega^2 P(t)\right) \tag{2.27}$$

Now, the slowly varying envelope approximation (SVEA) will be applied [HAK85], which uses the fact that the envelope function does not change significantly during one period T_{opt} of the fast optical oscillation,

$$|\dot{E}| \ll \omega|E| = \frac{2\pi}{T_{opt}}|E|, \tag{2.28}$$

so that only the terms of lowest order dominate Eq. (2.27) and the other can be neglected. The equation for the electric field amplitude E then is:

$$\frac{dE}{dt} = \frac{i\omega\Gamma}{2\epsilon_0\epsilon_{bg}}P, \tag{2.29}$$

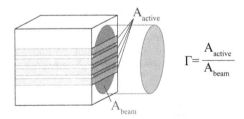

$$\Gamma = \frac{A_{active}}{A_{beam}}$$

Figure 2.6: Sketch of the confinement factor used in calculating the mode volume in a semiconductor laser. The confinement factor quantifies the overlap between electric field (red area) and active medium (blue layers).

where the confinement factor Γ was include. The confinement factor is a phenomenological addition and quantifies the fraction of the electric field overlapping with the gain medium in the laser (see Fig. 2.6), because in a semiconductor laser the extent of the standing electric field is usually bigger than the active region, i.e. the QD layer. Higher confinement leads to a more concentrated electric field profile and can be achieved by wave guiding. This results in a stronger interaction between the QDs and the light, but can also damage the semiconductor material, if intensities surpass the damage threshold of the material.

2.2.2. Matter Equations

After the field equations could be derived, the light-matter interaction and internal dynamics of the gain medium need to be described. To do so, it is necessary to derive the macroscopic polarization $\boldsymbol{P}$ as a function of the internal state of the gain medium [HAK85, CHO99]. On a fundamental level, the electric field is interacting with an optical transition, which needs to be inverted to facilitate lasing. This optical transition is in the most basic form a two-level system of electronic states, between which a transition is possible.

This can be described in the framework of quantum mechanics, of which the representation in second quantization will be used here. Furthermore, the electric field $\boldsymbol{E}$ is still described classically and not in the form of quantum electrodynamics. Without derivation, the Hamiltonian $\hat{H}$ for such a system is given by [CHO99]:

$$
\hat{H} = H_0 + H_s = \sum_{\alpha_j} \epsilon_{\alpha_j} a^\dagger_{\alpha_j} a_{\alpha_j} + \sum_{\beta_j} \epsilon_{\beta_j} b^\dagger_{\beta_j} b_{\beta_j}
$$
$$
- \sum_{\alpha_j, \beta_j} (\mu_{\alpha_j \beta_j} a^\dagger_{\alpha_j} b^\dagger_{\beta_j} + \mu^*_{\alpha_j \beta_j} a_{\alpha_j} b_{\beta_j}) \mathrm{Re}(E(t) e^{-i\omega t}). \tag{2.30}
$$

It consists of two parts, the single-state energies H_0 and the interaction H_s. α_j and β_j are sets of suitable quantum numbers for the upper and lower electronic levels, e.g. spin or wave number k. Then, a_{α_j} is the creation operator for an electron in the upper state and the hermitian conjugate $a^\dagger_{\alpha_j}$ the corresponding annihilation operator. Conversely, b_{β_j} and $b^\dagger_{\beta_j}$ are the creation and annihilation operator for holes in the lower state.

The number operators $a^\dagger_{\alpha_j} a_{\alpha_j}$ and $b^\dagger_{\beta_j} b_{\beta_j}$ count the number of electrons and holes, respectively. Together with the single particle state energies ϵ_{α_j} and ϵ_{β_j} the first two terms of Eq. (2.30) account for the energy of all occupied states. The last term describes the interaction with the electric field of amplitude E and frequency ω. Here, $\mu_{\alpha_j \beta_j}$ denotes the transition matrix element between state α_j and β_j and $\mu^*_{\alpha_j \beta_j}$ its complex conjugate.

From these quantum mechanic operators some observables can be derived [CHO99]. They are linked to the expectation value $\langle . \rangle$ and read:

$$
\rho_{e,\alpha_j} := \left\langle a^\dagger_{\alpha_j} a_{\alpha_j} \right\rangle \tag{2.31}
$$

$$
\rho_{h,\beta_j} := \left\langle b^\dagger_{\beta_j} b_{\beta_j} \right\rangle \tag{2.32}
$$

$$
\tilde{p}_{\alpha_j, \beta_j} := \langle b_{\beta_j} a_{\alpha_j} \rangle = \left\langle a^\dagger_{\alpha_j} b^\dagger_{\beta_j} \right\rangle^*, \tag{2.33}
$$

where ρ_{e,α_j} (ρ_{h,β_j}) is the average electron (hole) occupation probability in state α_j (β_j) and $\tilde{p}_{\alpha_j,\beta_j}$ is the microscopic dipole polarization amplitude for the optical transition α_j-β_j. The time evolution can now be obtained either in the Heisenberg representation of quantum mechanics or through the Ehrenfest theorem, for details see Ref. [SCU97]. The expectation value of an operator $\hat{o}$ then changes according to:

$$\frac{\partial}{\partial t} \langle \hat{o} \rangle = \frac{i}{\hbar} \left\langle \left[\hat{H}(t)\hat{o}(t) \right] \right\rangle = \hat{H}(t)\hat{o}(t) - \hat{o}(t)\hat{H}(t). \tag{2.34}$$

With [.] denoting the commutator as shown above. Hence, for the time evolution of the observables in Eq. (2.33) the commutator with the Hamiltonian has to be evaluated. This is a tedious calculation, which will not be shown here, where the anticommutator relation for fermions has to be used several times:

$$\left\{ a_m^\dagger, a_n \right\} := a_m^\dagger a_n + a_n a_m^\dagger = \delta_{mn}. \tag{2.35}$$

At the end, the following differential equations are obtained:

$$\frac{\mathrm{d}\tilde{p}_{\alpha_j \beta_j}}{\mathrm{d}t} = - i\omega_{\alpha_j \beta_j}\tilde{p}_{\alpha_j \beta_j} - \frac{i}{\hbar}\mu_{\alpha_j \beta_j}\mathrm{Re}\left(E(t)e^{-i\omega t}\right)\left(\rho_{e,\alpha_j} + \rho_{h,\beta_j} - 1\right) \tag{2.36}$$

$$\frac{\mathrm{d}\rho_{e,\alpha_j}}{\mathrm{d}t} = \frac{\mathrm{d}\rho_{h,\beta_j}}{\mathrm{d}t} = - \frac{i}{\hbar}\left(\mu_{\alpha_j \beta_j}^*\tilde{p}_{\alpha_j \beta_j}^* - \mu_{\alpha_j \beta_j}\tilde{p}_{\alpha_j \beta_j}\right)\mathrm{Re}\left(E(t)e^{-i\omega t}\right), \tag{2.37}$$

where $\omega_{\alpha_j \beta_j}$ is the frequency of the transition α_j-β_j. Then, the microscopic polarization $\tilde{p}_{\alpha_j \beta_j}$ amplitude can be transformed into a new variable $p_{\alpha_j \beta_j}$, describing the slowly varying amplitude in a rotating frame with the same frequency ω as the incoming electric field:

$$\tilde{p}_{\alpha_j \beta_j} =: p_{\alpha_j \beta_j}e^{-i\omega t}. \tag{2.38}$$

And the time evolution of this new variable is given by:

$$\frac{\mathrm{d}\tilde{p}_{\alpha_j \beta_j}}{\mathrm{d}t} = \frac{\mathrm{d}p_{\alpha_j \beta_j}}{\mathrm{d}t}e^{-i\omega t} - i\omega p_{\alpha_j \beta_j}e^{-i\omega t}. \tag{2.39}$$

This can now be used together with the expansion of $\mathrm{Re}\left(E(t)e^{-i\omega t}\right)$:

$$\mathrm{Re}\left(E(t)e^{-i\omega t}\right) = \frac{1}{2}\left(E(t)e^{-i\omega t} + E^*(t)e^{i\omega t}\right), \tag{2.40}$$

and inserting Eq. (2.39) and Eq. (2.40) into Eq. (2.37), while using the definition of $p_{\alpha_j \beta_j}$ yields:

$$\frac{\mathrm{d}p_{\alpha_j \beta_j}}{\mathrm{d}t} = - i\left(\omega_{\alpha_j \beta_j} - \omega\right)p_{\alpha_j \beta_j} - \frac{i}{2\hbar}\mu_{\alpha_j \beta_j}\left(E(t) + E^*(t)e^{i2\omega t}\right)\left(\rho_{e,\alpha_j} + \rho_{h,\beta_j} - 1\right) \tag{2.41}$$

$$\frac{\mathrm{d}\rho_{e,\alpha_j}}{\mathrm{d}t} = \frac{\mathrm{d}\rho_{h,\beta_j}}{\mathrm{d}t} = - \frac{i}{2\hbar}\left(\mu_{\alpha_j \beta_j}^*p_{\alpha_j \beta_j}^*e^{i\omega t} - \mu_{\alpha_j \beta_j}p_{\alpha_j \beta_j}e^{-i\omega t}\right)\left(E(t)e^{-i\omega t} + E^*(t)e^{i\omega t}\right). \tag{2.42}$$

These equations now contain terms oscillating with frequency 2ω. As the intrinsic time scales of the polarization amplitude $p(t)$ and electric field amplitude $E(t)$ are several orders of magnitude larger, these fast oscillating terms can be neglected. On these long time scales, they average out to zero. Additionally, the Rabi-frequency $\Omega_{\alpha_j\beta_j}$ is introduced as:

$$\Omega_{\alpha_j\beta_j}(t) = \frac{\mu_{\alpha_j\beta_j}E(t)}{\hbar}, \tag{2.43}$$

so that the semiconductor-Bloch-equations are finally derived:

$$\frac{dp_{\alpha_j\beta_j}}{dt} = -i\left(\omega_{\alpha_j\beta_j} - \omega\right)p_{\alpha_j\beta_j} - \frac{i\Omega_{\alpha_j\beta_j}}{2}\left(\rho_{e,\alpha_j} + \rho_{h,\beta_j} - 1\right) - \frac{1}{T_2}p_{\alpha_j\beta_j} \tag{2.44}$$

$$\frac{d\rho_{e,\alpha_j}}{dt} = \frac{d\rho_{h,\beta_j}}{dt} = -\operatorname{Im}\left(\Omega_{\alpha_j\beta_j}p_{\alpha_j\beta_j}^*\right) + R_{sp}^m + R_{scat}^m \tag{2.45}$$

Here, some additional phenomenological terms have been added. In addition to the coherent dynamics as derived from the Hamiltonian, the polarization amplitude p decays with the dephasing time T_2, accounting for the decay due to scattering processes such as carrier-carrier scattering [KOC00] and carrier-phonon scattering. Furthermore, the carrier densities ρ_{e,α_j} and ρ_{h,β_j} are subject to losses due to spontaneous emission R_{sp} and input and output R_{scat} by carrier scattering. The final forms for these terms used in this works will be shown later.

But first, the notation will be simplified to more closely match the two-state lasing quantum dots. Therefore, the indeces α_j and β_j will be replaced by $m \in \{GS, ES\}$, sorting everything into QD excited state (ES) and ground state (GS) variables. Furthermore, it will be assumed that only these two transitions are active, so that $\mu_{\alpha_j\beta_j} = 0$ and $p_{\alpha_j\beta_j} = 0$ for all other recombination processes. The notation then reads:

$$p^m := p_{\alpha_j\beta_j}$$
$$\rho_e^m := \rho_{e,\alpha_j}$$
$$\rho_h^m := \rho_{e,\beta_j}$$
$$\Omega^m := \Omega_{\alpha_j\beta_j} = \frac{\mu^m E^m(t)}{\hbar} \tag{2.46}$$

Now, the polarization dynamics in QD lasers are very fast [CHO99, BIM08a]. Hence, for the description on time-scales larger than the microscopic dephasing time, they can be adiabatically eliminated. It will therefore be assumed, that the polarization relaxes at all times to a value as given by the other variables. By setting the time evolution of the complex conjugated p^{m*} to zero,

$$0 = \frac{dp^{m*}}{dt} = \frac{i\Omega^{m*}}{2}\left(\rho_e^m + \rho_h^m - 1\right) - \frac{1}{T_2}p^{m*}, \tag{2.47}$$

and solving for the value of p^{m*},

$$p^{m*} = iT_2\Omega^{m*}\left(\rho_e^m + \rho_h^m - 1\right),$$ (2.48)

the semiconductor Bloch-equations can be reduced to the two differential equations for ρ^m and their final form can be obtained:

$$\frac{d\rho_b^m}{dt} = -\operatorname{Im}\left(\Omega^{m*}p^{m*}\right) + R_{sp} + R_{scat}$$

$$= T_2\frac{|\mu^m|^2}{\hbar^2}|E^m|^2\left(\rho_e^m + \rho_h^m - 1\right) + R_{sp} + R_{scat}$$ (2.49)

Finally, the result for the steady state of p^{m*} given in Eq. (2.48) can be used to calculate the macroscopic polarization P^m:

$$P^m(t) = \mu^m Z^{QD} p^m(t).$$ (2.50)

Here Z^{QD} denotes the number of quantum dots in the active medium, while μ^m is, as before, the microscopic polarization of the transition. The macroscopic polarization P^m is the dynamic variable that was missing in the description of the electric field dynamics given in Eq. (2.29). When inserting, the electric field equation is obtained as:

$$\frac{dE^m}{dt} = \frac{i\omega\Gamma}{2\epsilon_0\epsilon_{bg}}P^m$$

$$= \frac{\omega\Gamma T_2\nu_m Z^{QD}|\mu^m|^2}{2\epsilon_0\epsilon_{bg}\hbar}E^m\left(\rho_e^m + \rho_h^m - 1\right)$$

$$= gE^m\left(\rho_e^m + \rho_h^m - 1\right),$$ (2.51)

where the prefactor g denotes the optical gain. However, this only models the stimulated emission, and followingly the addiion of a spontaneous emission term will be covered in the next section.

2.2.3. Modelling of Spontaneous Emission

For the carrier occupation probabilities ρ_e^m and ρ_h^m, the loss by spontaneous recombination of electron-hole pairs can be modelled deterministically by using the Einstein-coefficients for spontaneous emission W_m:

$$R_{sp}^m = -W_m\rho_e^m\rho_h^m.$$ (2.52)

Followingly, photons are emitted at the same rate. But not all photons that are spontaneously emitted, are emitted in the right direction or polarization, so that many of them do not add to the number of photons N^{ph} in the lasing mode. Only the fraction β [HAK86] is emitted to further the lasing process, therefore Eq. (2.52) has to be modified accordingly for calculating the spontaneous emission contribution $\partial_t N_{sp}^{ph,m}$ to the photon number:

$$\partial_t N_{sp}^{ph,m} = \nu_m \beta Z^{QD} W_m \rho_e^m \rho_h^m. \tag{2.53}$$

Here, Z^{QD} denotes the number of quantum dots that are emitting and ν_m is the degeneracy of the level m. The rate of photon creation in Eq. (2.53) now has to be converted into electric field amplitude change. This is done by assuming energy conservation, for which the conversion between photons N^{ph} and electric field amplitude E is given by:

$$\frac{V_{mode}}{2}\epsilon_0\epsilon_{bg}\partial_t|E_{sp}|^2 = \hbar\omega\partial_t N_{sp}^{ph}, \tag{2.54}$$

where V_{mode} denotes the spatial extent of the electric field mode:

$$V_{\text{mode}} = \frac{Aha_l}{\Gamma}, \tag{2.55}$$

and A is the in-plane area of the active medium, e.g. the quantum dot layer, h is the height of one active medium layer, a_l is the number of layers and Γ is the confinement factor.

Accordingly, the change of energy in the mode has to be the same for both photons and electric field, so that:

$$\partial_t N_{sp}^{ph,m} = \beta\nu_m W_m \rho_e^m \rho_h^m = \frac{V_{mode}\epsilon_0\epsilon_{bg}}{2\hbar\omega}\partial_t|E_{sp}|^2, \tag{2.56}$$

where Eq. (2.53) was used. This can now be reshuffled to yield an expression for $\partial_t|E_{sp}|^2$:

$$\partial_t|E_{sp}|^2 = \nu_m\beta Z^{QD}W_{GS}\eta_{GS}^2\rho_e^{GS}\rho_h^{GS}, \tag{2.57}$$

describing the time evolution due to spontaneous emission of the amplitude square $|E_{sp}|^2$, where η_m^2 is the conversion factor between electric field amplitude and photon number N^{ph}, which is given by:

$$\eta_m^2 = \frac{(2\hbar\omega_m)}{(V_{\text{mode}}\epsilon_{bg}\epsilon_0)}. \tag{2.58}$$

In a last step, the derivative for the intensity $\partial_t|E|^2$ has to be converted into electric field amplitude change [FLU07]:

$$\partial_t E = \frac{E}{2|E|^2} \partial_t |E|^2 \tag{2.59}$$

Thus, the resulting spontaneous emission term is:

$$\partial_t E_{sp} = \beta Z^{QD} W_{GS} \eta_{GS}^2 \rho_e^{GS} \rho_h^{GS} \frac{E^{GS}}{|E^{GS}|^2}. \tag{2.60}$$

This can now be added to the field equation of Eq. (2.51). The complete set of rate equations will be presented in the next section.

2.3. Model of a Quantum Dot Laser

2.3.1. Dynamical Equations

After the full matter and electric field equations have been derived in the previous section, their final form shall be presented here. Although there exists a wide variety of more complex modelling approaches [CHO13], e.g. taking the k-distribution of carriers in the QW-reservoir into account [LIN11b], previous studies [LIN10] have shown that a simple rate equation approach is sufficient for the scope of this work. The numerical model used is based on previous works [MAJ11, LIN13, LIN14]. It includes the ground state (GS) and excited state (ES) variables for the electric field E^m, electron ρ_e^m and hole occupation probability ρ_h^m, with $m \in \{GS, ES\}$. The reservoir carrier densities w_e and w_h for holes and electrons are modelled as well.

In addition to the semiclassical laser equations derived from light-matter interaction of the previous section, some phenomenological terms have been added to account for additional processes, e.g. spontaneous emission, which will be explained in the following.

First, another feature of QDs must be taken into account: As of today no set of self-assembled QDs consists of identically shaped and sized QDs. Due to the nature of the growth process they exhibit a stochastic distribution of sizes, which in turn changes their confinements and spectral parameters [BIM08a]. This can be measured as a broadening of the collective spectrum, referred to as 'inhomogeneous broadening' (see Fig. 2.7), as compared to the natural linewidth of a single quantum dot (called 'homogeneous broadening'). Yet, only the spectral fraction of QDs with

Table 1: Dynamical variables of the numerical model

Symbol	Values	Meaning
$E^{[GS,ES]}$	$\in \mathbb{C}$	Slowly varying electric field amplitude for GS/ES
$\rho_{[e,h]}^{[ES,GS]}$	$0.0 - 1.0$	ES/GS electron/hole occupation probability for active QDs
$\rho_{[e,h],inact}^{[ES,GS]}$	$0.0 - 1.0$	ES/GS electron/hole occupation probability for inactive QDs
$w_{[e,h]}$	≥ 0	Electron/hole 2D-density in reservoir

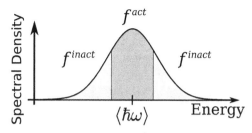

the highest density will end up lasing [BIM08a]. As a first and proven valid [LIN14] approximation, the QDs are divided into a fraction of f^{act} optically active and $1 - f^{act} = f_{inact}$ optically inactive QDs (coloured parts in Fig. 2.7). Therefore, a second set of occupations ($\rho^m_{e,inact}$ and $\rho^m_{h,inact}$) is included to account for optically inactive QDs.

The resulting equations for the slowly varying electric field amplitude are:

$$\frac{d}{dt}E^{GS} = \left[g_{GS}\left(\rho_e^{GS} + \rho_h^{GS} - 1\right) - \kappa\right]E^{GS}$$
$$+ \beta Z^{QD} f^{act} W_{GS}\eta^2_{GS}\rho_e^{GS}\rho_h^{GS}\frac{E^{GS}}{|E^{GS}|^2}, \qquad (2.61)$$

$$\frac{d}{dt}E^{ES} = \left[g_{ES}\left(\rho_e^{ES} + \rho_h^{ES} - 1\right) - \kappa\right]E^{ES}$$
$$+ \beta Z^{QD} f^{act} W_{ES}\eta^2_{ES}\rho_e^{ES}\rho_h^{ES}\frac{E^{ES}}{|E^{ES}|^2}. \qquad (2.62)$$

The first term accounts for the stimulated emission in accordance with the derivation in Sec. 2.1, where a decay term κE^m was added, accounting for the continuous loss of light intensity at the mirrors and by off-resonant absorption. The linear optical gain $g_{[GS,ES]}$ and decay rate κ are measured in units of $[1/ps]$. The second term deterministically models spontaneous emission. The number of QDs Z^{QD}, the Einstein-coefficient for spontaneous emission $W_{[GS,ES]}$ and the β-factor are without unit. The β-factor is equal to the fraction of spontaneously emitted photons that enter the lasing mode of the resonator, whereas the majority usually is lost.

The final form of the matter equations for the active quantum dots with $b \in \{e, h\}$ are:

$$\frac{d}{dt}\rho_b^{GS} = -\frac{g_{GS}}{\nu_{GS}Z^{QD}f^{act}}\left(\rho_e^{GS} + \rho_h^{GS} - 1\right)\frac{|E^{GS}|^2}{\eta^2_{GS}} - W_{GS}\rho_e^{GS}\rho_h^{GS}$$
$$+ S_{b,in}^{GS,cap}\left(1 - \rho_b^{GS}\right) - S_{b,out}^{GS,cap}\left(\rho_b^{GS}\right)$$
$$+ S_{b,in}^{rel}\left(1 - \rho_b^{GS}\right)\rho_b^{ES} - S_{b,out}^{rel}\rho_b^{GS}\left(1 - \rho_b^{ES}\right), \qquad (2.63)$$

$$\frac{d}{dt}\rho_b^{ES} = -\frac{g_{ES}}{\nu_{ES}Z^{QD}f_{act}}\left(\rho_e^{ES} + \rho_h^{ES} - 1\right)\frac{|E^{ES}|^2}{\eta_{ES}^2} - W_{ES}\rho_e^{ES}\rho_h^{ES}$$
$$+ S_{b,in}^{ES,cap}\left(1 - \rho_b^{ES}\right) - S_{b,out}^{ES,cap}\left(\rho_b^{ES}\right)$$
$$- \frac{1}{2}\left[S_{b,in}^{rel}\left(1 - \rho_b^{GS}\right)\rho_b^{ES} - S_{b,out}^{rel}\rho_b^{GS}\left(1 - \rho_b^{ES}\right)\right].\qquad(2.64)$$

Inactive QDs experience identical scattering rates, but lack any contribution by stimulated emission. They still emit spontaneously, though most of their emission light leaves the cavity quickly and is then lost:

$$\frac{d}{dt}\rho_{b,inact}^{GS} = -W_{GS}\rho_{e,inact}^{GS}\rho_{h,inact}^{GS}$$
$$+ S_{b,in}^{GS,cap}\left(1 - \rho_{b,inact}^{GS}\right) - S_{GS,out}^{GS,cap}\left(\rho_{b,inact}^{GS}\right)$$
$$+ \left[S_{b,in}^{rel}\left(1 - \rho_{b,inact}^{GS}\right)\rho_{b,inact}^{ES} - S_{b,out}^{rel}\rho_{b,inact}^{GS}\left(1 - \rho_{b,inact}^{ES}\right)\right],\qquad(2.65)$$

$$\frac{d}{dt}\rho_{b,inact}^{ES} = -W_{ES}\rho_{e,inact}^{ES}\rho_{h,inact}^{ES}$$
$$+ S_{b,in}^{ES,cap}\left(1 - \rho_{b,inact}^{ES}\right) - S_{b,out}^{ES,cap}\left(\rho_{b,inact}^{ES}\right)$$
$$- \frac{1}{2}\left[S_{b,in}^{rel}\left(1 - \rho_{b,inact}^{GS}\right)\rho_{b,inact}^{ES} - S_{b,out}^{rel}\rho_{b,inact}^{GS}\left(1 - \rho_{b,inact}^{ES}\right)\right].\qquad(2.66)$$

2D-reservoir variables w_e and w_h are measured in units of $[1/nm^2]$ and count charge carriers per area. The reservoir is filled with a constant influx of carriers by the current density J and decays at a rate of R_{loss}^W, which models all radiative and non-radiative loss processes. The scattering processes are weighted according to the area density of QDs N^{QD} given and the degeneracy of the levels (2 for GS, 4 for ES):

$$\frac{d}{dt}w_b = + J - R_{loss}^W w_e w_h \qquad(2.67)$$
$$- 2N^{QD}f^{act}\left[S_{b,in}^{GS,cap}(1 - \rho_b^{GS}) - S_{b,out}^{GS,cap}(\rho_b^{GS})\right]$$
$$- 4N^{QD}f^{act}\left[S_{b,in}^{ES,cap}(1 - \rho_b^{ES}) - S_{b,out}^{ES,cap}(\rho_b^{ES})\right]$$
$$- 2N^{QD}(1 - f^{act})\left[S_{b,in}^{GS,cap}(1 - \rho_{b,ia}^{GS}) - S_{b,out}^{GS,cap}(\rho_{b,ia}^{GS})\right]$$
$$- 4N^{QD}(1 - f^{act})\left[S_{b,in}^{ES,cap}(1 - \rho_{b,ia}^{ES}) - S_{b,out}^{ES,cap}(\rho_{b,ia}^{ES})\right],$$
$$\qquad(2.68)$$

For visualization, a sketch of the energy band structure and scattering processes of the QD model is shown in Fig. 2.8. The energy spacing for electrons and holes is

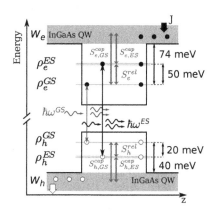

Figure 2.8: Sketch of the energy band structure, scattering processes and optical transition of the QD model. J denotes the pump current, ρ_b^m are occupation probabilities and w_b are the 2D-reservoir densities. There is an asymmetric energy spacing for electrons (e) and holes (h). The excited state (ES) has a degeneracy of two. ©(2015) IEEE. Reprinted, with permission, from [ROE14]

different, enabling asymmetric carrier dynamics. Injection current J only enters the system in the 2D-reservoir densities w_b. Recombination of GS electron-hole pairs and ES electron-hole pairs are the two lasing transitions described by the numerical model with E^{GS} and E^{ES}. Spin degeneracy is generally suppressed, but the ES level is assumed to be twice degenerate in comparison to the GS.

Table 2: Parameters for the QD model.

Symbol	Value	Meaning
$\hbar\omega_{GS}$	0.952eV	GS transition energy
$\hbar\omega_{ES}$	1.022eV	ES transition energy
a_L	15	number of QD layers
l	1mm	device length
d	2.4μm	device width
h	4nm	height of one layer
Γ	0.05	confinement factor
ϵ_{bg}	14.2	background permittivity
η_{GS}	9.157×10^{-7}V/nm	electric field conversion factor (GS)
η_{ES}	9.51×10^{-7}V/nm	electric field conversion factor (ES)

The optical transitions for the GS has an energy of $\hbar\omega_{GS} = 0.952$ eV, while the ES has $\hbar\omega_{ES} = 1.012$ eV. With the device assumed to have length $l = 1$ mm, width $d = 2.4$ μm and 15 active layers of height $h = 4$ nm, the mode volume can be calculated to $V_{\text{mode}} = 2.88 \times 10^{-15}$ m^3 for a confinement factor $\Gamma = 0.05$. The permittivity of GaAs is $\epsilon_{bg} = 14.2$, so that the conversion factors between photon number and electric field amplitude can be calculated. The results are given in Tab. 2, together with the parameters used in the calculation.

Now only the scattering rates need to be described, as is done in the next section.

2.3.2. Scattering Rates

Calculating the scattering rates for a semiconductor is a difficult task, as there are many possible recombination and transition processes in a semiconductor[SCH89]. The scattering rates S used in this work were calculated by a microscopic approach in previous works [MAL07, MAJ11, MAJ12]. The carrier-carrier interactions are modelled by a Born-Markov approximation, while the phonon-scattering is neglected. This leaves room for future improvements, as the phonon-assisted carrier capture might be especially important when an extra ES is included. A typical nonlinear fit of the scattering rates is given by:

$$S_{e,in}^{GS,cap} = \frac{(A_1 w_e^2 + A_2 w_h^2)\exp(C_1 w_e + C_2 w_h)}{1 - B_1 w_h/w_e + B_2 w_e + B_3 w_h - B_4 w_e^2 + B_5 w_h^2 + B_6 w_e w_h}. \tag{2.69}$$

The constants A_i, B_i and C_i are given in App. A.1. The corresponding out-scattering rate is calculated by the *detailed balance condition*: In thermal equilibrium, when in and out-scattering cancel each other, the occupation probabilities of the participating energy levels ρ_1 and ρ_2 are given by the Fermi-distribution:

$$\rho_1 = \frac{1}{\exp(\frac{E_1 - E_F}{k_b T}) + 1}, \tag{2.70}$$

$$\rho_2 = \frac{1}{\exp(\frac{E_2 - E_F}{k_b T}) + 1}, \tag{2.71}$$

where $E_2 > E_1$ are the potential energy of the levels, T is the temperature, k_B is the Boltzmann-constant and E_F is the Fermi-energy. The effective scattering terms in Eq. (2.63) and Eq. (2.64) have to cancel each other:

$$S_{in}(1 - \rho_1)\rho_2 = S_{out}(1 - \rho_2)\rho_1. \tag{2.72}$$

Inserting Eq. (2.71) and reshuffling yields the detailed balance condition in terms of scattering amplitude:

$$\frac{S_{out}}{S_{in}} = \frac{(1 - \rho_1)\rho_2}{(1 - \rho_2)\rho_1}$$
$$= \exp(-\frac{E_2 - E_1}{k_b T}), \tag{2.73}$$

and hence:

$$S_{out} = S_{in}\exp(-\frac{E_2 - E_1}{k_b T}) \tag{2.74}$$

So that only the energy difference $E_2 - E_1$ enters the detailed balance condition, not the absolute energy. Out-scattering is faster for higher temperatures T and smaller energy spacing, i.e. weaker confinement.

Furthermore, QDs of different sizes will later be investigated. To study them a size scaling parameter r will be introduced and scattering rates are calculated for all of the different QD ensembles. Because of computational limitations, only a linearised fit of the microscopically calculated scattering rates is used. This linearised fit is given in App. A.2. Detailed balance, on the other hand, is fully maintained.

3. Modes of Operation of QD Lasers

3.1. Single Colour Laser

Before delving into the dual-colour or two-state lasing aspects of the QD laser model introduced in the previous section, this section will shortly review the single-colour dynamics of a QD laser. This is done by setting the ES gain $g_{ES} = 0$, so that only GS lasing is achieved. Then, the ES simply acts as an intermediate reservoir for the carriers.

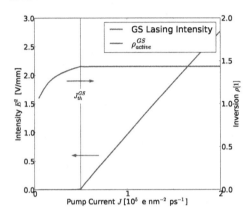

Figure 3.1: Light-current characteristic for a single colour lasing QD laser as simulated by the numerical model introduced in section Sec. 2.3 with parameters as given in Tab. 3. The vertical line marks the lasing threshold J_{th}^{GS}. After the onset GS lasing intensity (red solid line), the GS occupation $\rho_{active}^{GS} = \rho_e^{GS} + \rho_h^{GS}$ (green solid line) is clamped.

Figure 3.1 shows the light-current characteristic, i.e. the light intensity E^2 versus injection current J of such a device. The parameters used for numerical simulation are derived from previous works [MAJ11, PAU12, LIN12a, LIN13] and are given in Tab. 3. The resulting GS intensity (red line) exhibits a clear lasing threshold J_{th}^{GS}, which is marked by the vertical line. This is similar to the results of the rate equation approach presented in Sec. 2.1.3. Additionally, the green line in Fig. 3.1 denotes the GS occupation of active dots $\rho_{active}^{GS} = \rho_e^{GS} + \rho_h^{GS}$. First, the occupation is rising with injection current J, but then saturates for values above the lasing threshold J_{th}^{GS}. This is the already mentioned *gain clamping* and a fundamental feature of lasing, caused by the interaction of carriers and electric field through stimulated emission. Here, ρ_{active}^{GS} corresponds to the inversion D of the simple two-variable model of Sec. 2.1.3.

The value of the inversion gain clamping can also be easily calculated by setting the Eq. (2.61) for the time evolution of the GS electric field amplitude E^{GS} to zero. The spontaneous emission term can be neglected, due to its small contribution during lasing operation, so that the following steady-state condition has to be solved:

$$0 = \left[g_{GS} \left(\rho_e^{GS} + \rho_h^{GS} - 1 \right) - \kappa \right] E^{GS} \tag{3.1}$$

Obviously, there are at least two solutions for the set of dynamic variables ρ_e^{GS}, ρ_h^{GS} and E^{GS}, similar to the simple rate equation model presented in Sec. 2.1.3: The

Table 3: Parameters used in the calculations for the single colour laser.

Symbol	Value	Meaning
T	300K	temperature
g_{GS}	0.115ps^{-1}	GS linear gain
g_{ES}	0.0ps^{-1}	ES linear gain
κ	0.05ps^{-1}	optical losses
β	2.2×10^{-3}	spontaneous emission factor
Z^{QD}	1.5×10^7	number of QDs
N^{QD}	$0.5 \times 10^3\text{nm}^2$	area density of QDs
f^{act}	0.366	fraction of active dots
W_{GS}	$4.4 \times 10^{-4}\text{ps}^{-1}$	GS spontaneous emission rate
W_{ES}	$5.5 \times 10^{-4}\text{ps}^{-1}$	ES spontaneous emission rate
R^W_{loss}	$0.09\text{nm}^2\text{ps}^{-1}$	QW loss rate
η_{GS}	$9.157 \times 10^{-7}\text{V/nm}$	electric field conversion factor (GS)
η_{ES}	$9.51 \times 10^{-7}\text{V/nm}$	electric field conversion factor (ES)

trivial *off*-state for $E^{GS} = 0$ and a second solution if the term in brackets equals zero:

$$0 = \left[g_{GS} \left(\rho_e^{GS} + \rho_h^{GS} - 1 \right) - \kappa \right]$$
$$\rho_e^{GS} + \rho_h^{GS} = \frac{\kappa}{g_{GS}} + 1. \tag{3.2}$$

This is also the solution, which allows a non-zero electric field amplitude E^{GS} and therefore corresponds to the lasing state or *on*-state. With the use of the active dot inversion definition $\rho_{active}^{GS} = \rho_e^{GS} + \rho_h^{GS}$ and the parameters in Tab. 3, the gain clamping can be calculated to lead to a saturation of inversion for $\rho_{active}^{GS} \simeq 1.43$ for the single colour laser modelled in this section. This result is confirmed by the steady state inversion above threshold in Fig. 3.1, where this value is reached.

Equation (3.2) for the gain clamping will be used several times throughout this work, so a closer look to some of its implications will be discussed here. First, it is obvious that the ratio of κ to g_{GS} can reach a wide variety of values, e.g. $\kappa/g_{GS} \simeq 0$ and $\kappa/g_{GS} \simeq 10$ are both obtainable through choice of mirror and gain properties. However, the inversion ρ_{active}^{GS} never falls below zero or goes beyond two, as it is the sum of two occupation probabilities $\rho_{active}^{GS} = \rho_e^{GS} + \rho_h^{GS}$. So for optical loss rates greater than the linear GS gain $\kappa/g_{GS} > 1$, the *on*-state is never reached. This corresponds to a laser with too low gain or too high losses, so that no continuous wave operation is possible. Conversely, for high-quality cavities with low κ, lasing is reached as soon as the inversion equals one.

Second, even though the inversion ρ_{active}^{GS} is clamped to a certain value, the individual contributions of ρ_e^{GS} and ρ_h^{GS} can vary, i.e. the ratio of electrons and holes is not determined. This is a feature of the non-excitonic approach taken in this work and will be of importance later.

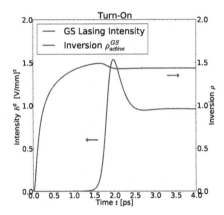

Figure 3.2: Turn-on time series for the single colour laser for $J = 1 \cdot 10^{-5} \mathrm{enm}^{-2}\mathrm{ps}^{-1} \simeq 2J_{th}^{GS}$. Parameters are given in Tab. 3. The GS inversion $\rho_{active}^{GS} = \rho_e^{GS} + \rho_h^{GS}$ (green line) starts to rise almost instantaneously, while GS intensity $|E^{GS}|^2$ (red line) exhibits a turn-on delay. Also visible are the highly damped relaxation oscillations and the steady state value for $\rho_{active}^{GS} \simeq 1.43$.

Figure 3.2 shows a turn-on time series of the single colour laser. The parameters were as given in Tab. 3 with $J = 1 \cdot 10^{-5}\mathrm{enm}^{-2}\mathrm{ps}^{-1}$, which is roughly $2J_{th}^{GS}$ as can be seen in Fig. 3.1. After the inversion (green line) rises for the first 1.5 ns, the GS lasing intensity $|E^{GS}|^2$ (red line) also turns on. This delay is caused by the reliance on stimulated emission, so that the intensity can only rise after the inversion reaches the gain clamping threshold value (see Eq.(3.2)). In comparison to the turn-on of the more simple two-variable rate equation system in Fig. 2.5 on page 13, the relaxation oscillations are highly damped, as is generally the case for QD lasers [LUE09a].

3.2. Two-State Lasing

3.2.1. Experiments and Interpretation

As mentioned in Sec. 1, in general lasing occurs by the recombination of electrons and holes in the ground state (GS) of the QD, lying at the intersection of a p-n junction. When the recombination is of charge carriers in excited state (ES) levels, lasing occurs at lower wavelengths [LUE12]. Simultaneous ES and GS lasing, called two-state lasing, is also possible [GRU97] and has been experimentally observed [MAR03a, SUG05b] for different QD materials (see Fig. 3 in [MAR03a]). Furthermore, lasing of the second excited state has been observed experimentally in three-state lasing devices(see Fig. 2 in [ZHA10a]).

The existence of two-state lasing devices is a proof of the finite time carrier dynamics and an *incomplete gain clamping* [MAR03c]. These finite time carrier processes are a special attribute of QDs and usually considered a disadvantage [BIM08a]. To understand the significance of this incomplete gain clamping, one has to first take a closer look at gain clamping in multi-level systems.

When the GS starts lasing, its carriers become clamped according to the gain clamping condition as introduced in the previous section:

$$\rho_e^{GS} + \rho_h^{GS} = \frac{\kappa}{g_{GS}} + 1. \qquad (3.3)$$

If one now assumes that the additional energetic levels in the system are linked to this optically active GS by fast scattering processes, as is the case in a QW laser, both the ES occupations ρ_b^{ES} and reservoir densities w_b will be determined by a Fermi distribution corresponding to the gain clamped GS occupation. Followingly, the ES will be less occupied than the GS, and the reservoir will be even emptier. Any further increase of the carrier input, e.g. by raising the pump current J, will be instantly redistributed throughout all levels. But the GS inversion cannot increase further as any additional carriers added will be converted into GS lasing light and the other populations are linked to this clamped GS inversions, so that in the end no carrier population can rise. For fast scattering rates, the carrier population as a whole is gain clamped.

However, if one includes a scattering rate that is slow, as is the case for a QD, the situation changes. As the carriers are injected into the QW and only the GS level is clamped, they accumulate in the ES and reservoir. This is of course dependent on the pump current J, but the ES can reach a sufficient inversion for lasing if enough carriers are injected. ES and reservoir dynamics are still subject to further dynamics, even after the onset of GS lasing. Thus, in this case the carrier population as a whole is not gain clamped.

Hence, a simultaneous emission on two different emission lines would be impossible to achieve for quasi-instantaneous equilibration of carrier populations, as is the case for QW lasers, so that the appearance of a second lasing line in the experiments is a clear indicator for the finite-time scattering processes. It especially reveals that *ES occupations cannot be inferred from GS occupations*, as the former is increasing despite the latter being gain clamped. This is important to keep in mind, as in the scope of this work it will later be shown that the reverse is possible.

3.2.2. Numerical Simulations

To verify these claims the numerical QD laser model of Sec. 2.3 is used for the simulation of two-state lasing. Figure 3.3 shows the light-current characteristic of a two-state laser for the parameters as given in Tab. 4. After the appearance of the GS emission (red line) for the GS threshold current J_{th}^{GS}, the ES also begins lasing (blue dotted line). The GS occupation $\rho_{active}^{GS} = \rho_e^{GS} + \rho_h^{GS}$ is clamped after the onset of GS lasing, as can be seen by the horizontal course of the green solid line. However, the microscopically calculated scattering rates used lead to an incomplete gain clamping for the rest of the system. Therefore, ES occupation $\rho_{active}^{ES} = \rho_e^{ES} + \rho_h^{ES}$ (green dotted line) is still increasing for increasing J and subsequently reaches sufficient levels to facilitate ES lasing.

This is in very good agreement with the explanation given by Markus et. al. (2003) of 'incomplete gain clamping' and the LI-curve looks sufficiently similar to be satisfactory. Now, however, as is only possible in numerical simulation, the opposite shall also be investigated. Fig. 3.4 shows a simulation with the same

Table 4: Parameters used in the calculations unless noted otherwise.

Symbol	Value	Meaning
T	300K	temperature
g_{GS}	0.115ps^{-1}	GS linear gain
g_{ES}	0.23ps^{-1}	ES linear gain
κ	0.05ps^{-1}	optical losses
ΔE_e	50meV	ES-GS energy gap for electrons
ΔE_h	20meV	ES-GS energy gap for holes
β	2.2×10^{-3}	spontaneous emission factor
Z^{QD}	1.5×10^7	number of QDs
N^{QD}	$0.5 \times 10^3\text{nm}^2$	area density of QDs
f^{act}	0.366	fraction of active dots
W_{GS}	$4.4 \times 10^{-4}\text{ps}^{-1}$	GS spontaneous emission rate
W_{ES}	$5.5 \times 10^{-4}\text{ps}^{-1}$	ES spontaneous emission rate
R^W_{loss}	$0.09\text{nm}^2\text{ps}^{-1}$	QW loss rate
η_{GS}	$9.157 \times 10^{-7}\text{V/nm}$	electric field conversion factor (GS)
η_{ES}	$9.51 \times 10^{-7}\text{V/nm}$	electric field conversion factor (ES)

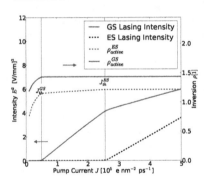

Figure 3.3: Light-current characteristic for a two-state lasing QD laser as simulated by the numerical model introduced in section Sec. 2.3 with parameters as given in Tab. 4. After the onset GS lasing intensity (red solid line), the GS occupation $\rho^{GS}_{active} = \rho^{GS}_e + \rho^{GS}_h$ (green solid line) is clamped. Yet, the ES occupation ρ^{ES}_{active} (green dashed line) is still rising, so that ES lasing (blue solid line) appears and two-state lasing is achieved.

parameters as Fig 3.3, but scattering rates increased by a factor of 100. The now nearly instantaneous scattering suppresses any increase of the ES occupation ρ^{ES}_{active} (green dashed line) after the GS lasing threshold J^{GS}_{th} and the system is fully gain clamped. Hence, no ES lasing can appear.

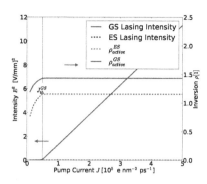

Figure 3.4: Light-Current Characteristic for a two-state lasing QD laser as simulated by the numerical model introduced in section Sec. 2.3 with parameters as given in Tab. 4, but scattering rates speed up by a factor of 100. After the onset GS lasing intensity (red solid line), the GS occupation $\rho_{active}^{GS} = \rho_e^{GS} + \rho_h^{GS}$ (green solid line) and the ES occupation ρ_{active}^{ES} (green dashed line) are clamped. Therefore no ES lasing appears and two-state lasing is not achieved as the system is *fully gain clamped*.

3.3. Ground State Quenching

3.3.1. Experiments and Description

In the first experimental measurement of two-state lasing by Markus *et al.* [MAR03a] some of the devices exhibited a complete roll-over of GS intensity for increasing currents (see Fig. 3 in [MAR03a], left panel or the schematic of Fig. 3.5). Since then several experimental observations have been published (e.g. see Fig. 3 in [MAX13]) and have been the topic of a debate in the literature [KOR13a, GIO12]. This paradoxical behaviour of a *decreasing* intensity for an *increasing* current has been termed 'GS quenching' and a description of its features known so far follows.

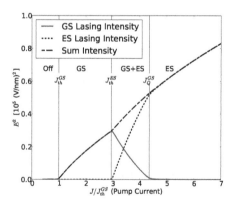

Figure 3.5: Simulated light-current characteristic for QDs exhibiting GS quenching. The GS intensity starts to decline after the onset of ES lasing, while the slope of the overall intensity stays roughly the same. ©(2015) IEEE. Reprinted, with permission, from [ROE14]

First, while two-state lasing has also been observed for InAs/InP QDs [VES07], GS quenching is unique to self-assembled InAs/GaAs QDs as far as the author knows. However, this is also caused by the small numbers of papers published on this topic. From a pure performance perspective GS quenching might seem like a defect, so some additional experimental observation might have been discarded or not deemed publishable.

Secondly, the principal course is always comparable: After the onset of ES lasing at the ES lasing threshold J_{th}^{ES} the GS intensity starts to decrease, while the overall intensity (GS + ES) increases further. The slope of the LI-curve is almost not affected. This is an indicator that the numbers of QDs participating in lasing stays constant throughout the entire current range. Otherwise an increase in the slope could be expected, as the differential gain is higher when more QDs are used. For the numerical model of this work introduced in Sec. 2.3 this means that the active fraction f^{act} of QDs (see Fig. 2.7) is composed of the same QDs in the GS and ES spectral range, e.g. there are no purely ES lasing QDs.

Lastly, GS quenching is dependent on device length as already studied by Markus et al. (see Fig. 3 in [MAR03a]). They found a critical length ℓ, below which only ES lasing was present, intermediate lengths with two-state lasing, and an increasing threshold current for the ES lasing for larger devices. This has also been independently confirmed in [CAO09] and [LEE11c] and the need of short cavities is also mentioned in [VIK07a]. Furthermore, a study of lasing thresholds by Maximov et al. [MAX13] has shown a background temperature and doping dependence.

Additional claims have been put forth, that GS quenching is absent in pulsed mode [LEE11c], while others [JI10, VIK05] have observed a GS quenching even in pulsed mode, though with increased thresholds. A further experimental investigation of pulsed mode GS quenching is therefore required.

3.3.2. Mechanisms of Quenching in the Literature

Throughout the literature several attempts have been made to explain and model this behaviour. They can be divided into three different approaches: self-heating, increase of homogeneous broadening and electron-hole dynamics. This section will introduce all three of them and compare them to available experimental evidence.

Self-heating, i.e. a pump-dependent T, was proposed by [MAR03c, JI10]. Due to Joule heating the device will run at elevated temperatures in the active region, while scattering processes may raise the temperature of the electron gas even further and thus lead to a broadening of the Fermi-function of the equilibrium occupation probability. This directly translates to faster out-scattering rates, caused by the *detailed balance condition* (see Eq. 2.73). Because of its higher energy, the excited state will usually be less occupied than the GS, but this difference will become less pronounced and vanish in the limit of very high temperatures. Caused by its higher degeneracy [LUE12], the ES gain g_{ES} is greater than the GS gain g_{GS}. When both levels have similar occupation numbers, the ES will therefore be the only state left lasing. The GS occupation probabilities will be closely tied to the ES level, but unable to reach sufficient inversion levels due to gain clamping.

This is in accordance with the observation in Ref. [LEE11c] that the GS does not quench in pulsed mode, which the authors attributed to the lack of self-heating. On the contrary, other sources report a GS quenching even while pulsing and either refute the self-heating theory altogether [ZHU12a, VIK05] or extend it to the pulsed regime as well [JI10]. Self-heating as the source of GS quenching is therefore still controversial. Note, however, that self-heating should also lead to GS quenching in InAs/InP-based two-state lasing QDs, where it has not been observed so far.

The second mechanism was introduced by Sugawara *et al.* [SUG05b]. The authors included the effects of a current dependent homogeneous broadening to model a decrease of the maximum gain. Increasing the homogeneous broadening leads to a decrease in the peak gain at the resonance frequency, thus for a high enough broadening GS lasing becomes impossible. They could fit their experimental results well, yet, as has already been pointed out [VIK05, GIO12, KOR13a], this effect may not be as strong as was assumed by Sugawara *et al.* [SUG05b]. Additionally, this effect should again also be applicable to InAs/InP-based QDs, where GS quenching has not been observed [VES07, PLA05].

Lastly, Viktorov *et al.* focus on the different energy separation factors and different transport time scales for holes and electrons [VIK05]. The asymmetric dynamics lead to a competition for holes between GS and ES, and when hole occupation is low, i.e. hole levels are depleted, the ES will win because of its higher degeneracy. The same approach is also applied in the model of Gioannini [GIO12] and the analytical solutions of Korenev *et al.* [KOR13, KOR13a].

This electron-hole asymmetry is highly dependent on the different energy spacing and scattering dynamics for electrons and holes inside the QDs, but leads to GS quenching when hole occupations are low. Thus, there must be a mechanism explaining how the electron to hole fraction increases with pump current. GS quenching can only appear if there was GS lasing to begin with. So there needs to exist a transition to a state of hole depletion from a state where enough holes could be supplied to facilitate GS lasing. Holes therefore have to get scarcer with increasing pump current and cannot be scarce all the time. Other authors have either neglected charge conservation [KOR13] and directly varied the capture rates to achieve this, or added other energy levels for the holes to accumulate in outside of the lasing states [GIO12].

Table 5: Numerical models used to study GS quenching. (SH = self-heating; HB = homogeneous broadening; EH = electron-hole asymmetry)

Author	excitonic	charge conserv.	# of subgroups	Quenching mechanism
Markus *et al.* (2003) [MAR03c]	Yes	Yes	1	SH
Sugawara *et al.* (2005) [SUG05b]	Yes	Yes	801	HB
Viktorov *et al.* (2005) [VIK05]	No	No	1	EH
Kim *et al.* (2010) [KIM10f]	No	Yes	Several	SH+HB
Ji *et al.* (2010) [JI10]	Yes	Yes	1	SH
Gioannini (2012) [GIO12]	No	Yes	2	EH
Korenev *et al.* (2013) [KOR13a]	No	No	1	EH
This work [ROE14]	No	Yes	2	EH

Since GS quenching and the above mentioned mechanisms touch on central topics like heating, energy spacings and homogeneous broadening, but are also, as will be shown, influenced by scattering rates, they deserve to be studied more deeply. Even after nearly two decades of QD research many fundamental questions are still open. Understanding GS quenching is therefore helpful in constraining parameters and gaining deeper insight into the non-excitonic carrier dynamics.

Table 5 lists numerical simulations of GS quenching and their attributes. Note that electron-hole asymmetry is not reproduceable with excitonic models.

Section 4.1 will present a novel analytical approach to study all of these mechanisms in a unified framework. But for the numerical part, electron-hole asymmetry is the mechanism leading to GS quenching in this work. This is due to the authors firm believe, that it's mainly the scattering dynamics and energy spacings that set apart the InAs/InP and InAs/GaAs material systems. The theory for the homogeneous broadening increase has not seen any further theoretical development and lacks a motivation from first principles, while self-heating is assumed to only play a minor role in GS quenching.

4. Understanding QD Laser Regimes of Operation

4.1. Analytical Approximations

4.1.1. Derivation

Understanding GS-quenching is made difficult, by the high dimensionality of the system and the many experimentally not accessible variables. When described in a non-excitonic picture, one needs to at least include four different carrier reservoirs and two electric field amplitudes. Korenev et al. [KOR13a] have analytically solved a somewhat reduced representation of this system, by assuming a common hole GS level and neglecting charge conservation. They are able to derive GS-quenching light-current characteristics by assuming that holes are less likely to enter the QD than electrons. Yet they lack an explicit modelling of the current dependence.

The model used in the scope of this work contains six carrier levels and includes microscopically motivated scattering rates, which allows the realistic modelling of current dependent carrier dynamics. However, general analytical solutions, even of the steady states, do not exist. Nonetheless, an analytical approximation shall be derived in this section, which visualizes the different explanations given in the previous section. A quantitative discussion of the order of magnitude of these effects is also possible, once some general assumptions about the device are made.

The GS occupation will now be derived as the equilibrium occupation as given by the ES occupations for vanishing stimulated emission. Assuming that the carriers in the GS are mainly dominated by relaxation from the ES, all but these terms can be neglected in the time evolution of Eq. (2.63):

$$\frac{d}{dt}\rho_b^{GS} = S_{in,b}^{rel}\rho_b^{ES}\left(1 - \rho_b^{GS}\right) - S_{out,b}^{rel}\rho_b^{GS}\left(1 - \rho_b^{ES}\right). \tag{4.1}$$

This assumption is valid for the scattering rates as derived for self-assembled InGaAs-QDs in [LIN13, LIN14] as used for this work, as direct capture into the GS is slower than the cascade scattering from reservoir to ES and then into the GS. However, this assumption is not obviously valid in general for other material systems. Furthermore, a different electronic structure even if fabricated from InGaAs, e.g. nanorods or asymmetric QDs, could also potentially invalidate this precondition.

Solving Eq. (4.1) for the steady state of $\frac{d}{dt}\rho_b^{GS} = 0$ yields:

$$\rho_b^{GS} = \frac{1}{1 + \left(\frac{1}{\rho_b^{ES}} - 1\right)\frac{S_{out,b}^{rel}}{S_{in,b}^{rel}}}, \tag{4.2}$$

The ratio of in to out scattering rates can be determined by the detailed balanced relation of Eq. (2.73):

$$\frac{S_{out,b}^{rel}}{S_{in,b}^{rel}} = e^{-\frac{\Delta E_b}{k_b T}}, \tag{4.3}$$

Here, ΔE_b is the energy difference between GS and ES for $b \in e, h$. Hence, the GS occupation is given by:

$$\rho_b^{GS} = \frac{1}{1 + \left(\frac{1}{\rho_b^{ES}} - 1\right) e^{-\frac{\Delta E_b}{k_b T}}}, \tag{4.4}$$

by which the system can be reduced to the excited state occupations ρ_b^{ES}. Because the stimulated emission terms of Eq. (2.73) have been neglected, Eq. (4.4) can yield GS occupations that are above the GS gain clamping, which is unphysical. These can be interpreted as GS lasing states, so that the gain clamping equation turns into a threshold conditions and the lasing condition is therefore given by:

$$\rho_e^{GS} + \rho_h^{GS} - 1 \geq \frac{\kappa}{g_{GS}}. \tag{4.5}$$

After inserting the analytical approximation of Eq. (4.4) into Eq. (4.5) and reshuffling, the lengthy equation

$$\rho_h^{ES} \geq \frac{(g_{GS} + \kappa)\left(1 - \rho_e^{ES}\right) + \rho_e^{ES}\kappa e^{\frac{\Delta E_e}{k_b T}}}{(g_{GS} + \kappa)\left(1 - \rho_e^{ES}\right) + (g_{GS} - \kappa)\rho_e^{ES} e^{-\frac{\Delta E_e + \Delta E_h}{k_b T}} + \kappa \rho_e^{ES}\left(e^{\frac{\Delta E_e}{k_b T}} + e^{\frac{\Delta E_h}{k_b T}}\right) - \kappa e^{\frac{\Delta E_h}{k_b T}}} \tag{4.6}$$

is obtained. This Eq. (4.6) expresses the lasing condition for the GS in terms of ES occupations. It is only valid for steady states, when the GS reservoirs have equilibrated with their ES counterparts. It only depends on a few key parameters: Linear gain g_{GS}, optical decay rate κ, energy spacing between levels ΔE_b and thermal energy $k_b T$. Their influence and meaning shall be discussed in the following section. By assuming that *GS occupations can be inferred from ES occupations*, the dimensionality of the system was reduced.

Lastly, expressing the ES gain clamping in terms of ES occupations is trivial, but enables the description of both ES and GS lasing thresholds in the plane of ES occupation probabilities ρ_b^{ES}. Hence, the ES is lasing if ES occupations reach sufficient values:

$$\rho_e^{ES} + \rho_h^{ES} - 1 = \frac{\kappa}{g_{ES}}. \tag{4.7}$$

4.1.2. Parameter Dependent Lasing Thresholds

Analytical expressions for both the ES gain clamping (Eq. (4.7)) and the GS lasing threshold (Eq. (4.5)) have now been obtained. For the parameters as given in Tab. 6 they are plotted in Fig. 4.1. Here, the x-axis is the ES electron occupation probability ρ_e^{ES}, while the y-axis is ES hole occupation probability ρ_h^{ES}. Yellow marks the

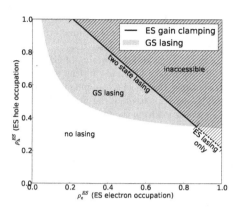

Figure 4.1: ES gain clamping and GS lasing regime vs. excited state electron and hole occupations. When the ES is lasing, the inversion is clamped at $\rho_e^{ES} + \rho_h^{ES} - 1 = \kappa/g_{ES}$ (Eq. (4.7), black line). Calculating the GS occupation probabilities through instant quasi-equilibrium yields the condition for GS lasing expressed in terms of ES occupations (Eq. (4.5), yellow area). The overlap of both corresponds to two-state lasing. Furthermore, low occupations lead to no lasing (white area). The ES occupation can never exceed ES gain clamping, hence the shaded area is inaccessible as a steady state solution. Parameters given in Tab. 6.

Table 6: Parameters used in the calculations of this section unless noted otherwise.

Symbol	Value	Meaning
T	300K	temperature
g_{GS}	0.115ps^{-1}	GS linear gain
g_{ES}	0.23ps^{-1}	ES linear gain
κ	0.05ps^{-1}	optical losses
ΔE_e	70meV	ES-GS energy gap for electrons
ΔE_h	10meV	ES-GS energy gap for holes

region where Eq. (4.5) is fulfilled and GS lasing is apparent, from here on called the 'GS lasing regime', whereas similarly the black line represents the limit of ES gain clamping (Eq. (4.5)), denoted as 'ES lasing regime'.

Thus, when reservoir densities are not relevant and the light intensity for ES and GS is only distinguished between lasing and non-lasing, the state of the system can be entirely represented by a point in ρ^{ES}-phase space. Generally, ES-occupations can lie anywhere between zero and one, but are additionally bounded by the ES gain-clamping line. The shaded area of Fig 4.1 is therefore inaccessible. The region where the GS lasing regime and the ES lasing regime intersect, i.e. where the black line lies inside the yellow area, two-state lasing is present.

However, despite this helpful visualisation of two-state lasing, current-dependent steady states are not calculable in this semi-analytical approach. They have to be calculated separately and can then be compared to the analytical approximations and boundaries in the ρ_b^{ES}-plane as later done in Sec. 4.2. But studying the extent and parameter dependence of the different lasing regimes can nevertheless yield valuable insights for the underlying mechanics of GS quenching.

Generally, Eq. (4.6) suggests that the position of the GS lasing regime and the extent of the overlap region indicating two-state lasing can be changed by changing the gain g or losses κ, the temperature T or the energy structure $\Delta E_e/\Delta E_h$. This is reflected in some of the GS quenching mechanisms already suggested in the literature, e.g. self-heating or gain suppression by homogeneous broadening increase, but has not been discussed coherently. Therefore, a variety of parameter sets and the implications for two-state lasing shall be discussed in this section.

For the set of parameters as taken in Fig. 4.1, there is an overlap of GS and ES-lasing regimes. This 'two-state lasing regime' marks the dual-emitting state of the QD laser. Additionally, for low hole occupations, the ES gain clamping is outside the GS lasing regime bounds and solitary ES lasing is apparent. If the current dependent scattering dynamics of the laser lead to a transition from the two-state lasing regime to the solitary ES lasing regime, a quenching of the GS can be observed.

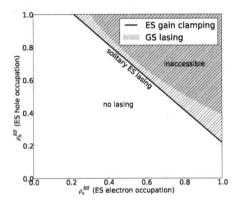

Figure 4.2: ES gain clamping and GS-lasing regime vs. ES electron and hole occupations for lower confinement ΔE_e = 15meV and ΔE_h = 5meV (other parameters given in Tab. 6). The ES gain clamping line (Eq. (4.7), black line) lies entirely outside of the GS-lasing regime (Eq. (4.5), yellow area). There is no overlap and therefore no two-state lasing regime.

The two-state lasing regime can be entirely absent for different parameters, when the GS lasing regime lies at too high ρ^{ES} to facilitate GS lasing. Figure 4.2 shows the analytical lasing regimes for lower electric confinement ΔE_e = 15meV and ΔE_h = 5meV. Here, the GS lasing regime lies entirely in the inaccessible part above the ES gain clamping line (shaded areas). This corresponds to a QD structure, where charge carriers are easily escaping the GS and both confined states have similar occupations. The ES will then be left lasing, because its higher degeneracy leads to a larger optical gain. As can be seen, two-state lasing is impossible for such a device, independent of the actual current-dependent dynamics, as the GS never turns on.

Consequently, increasing the confinement to ΔE_h = 30meV will prevent any solitary ES lasing. As seen in Fig. 4.3 the GS-lasing (yellow area) regime is covering the entire extent of the ES gain clamping line (black), which means there is no steady state of purely ES lasing. The system can and possibly will traverse the GS lasing-regime for increasing currents and end up in a two-state lasing regime on the gain clamping limit, but it has nowhere to go from there.

Note that the parameters given in Tab. 6 and the follow up examples all feature significantly smaller energetic hole confinement ΔE_h than electron confinement ΔE_e. On the one hand this is based on the microscopically calculated electronic structure of real QD devices [SCH07f], but on the other hand this has emerged in the scope of this analytical approach as a key feature of lasers exhibiting a GS-quenching

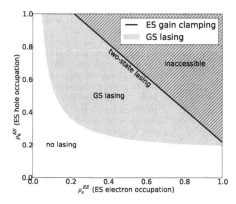

Figure 4.3: ES gain clamping and GS lasing regime vs. excited state electron and hole occupations for stronger hole confinement $\Delta E_h = 30\text{meV}$ (other parameters given in Tab. 6). No solitary ES-lasing present, as the ES gain-clamping line (black) lies entirely inside the GS-lasing regime (yellow area).

behaviour. This is in accordance with the work of Viktorov et al. [VIK05], where they proposed an asymmetry-based GS-quenching mechanism.

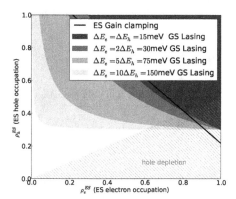

Figure 4.4: Energy asymmetry effects on GS quenching. When GS and ES energy separation is the same for holes and electrons, the ES inversion clamping (black line) is intersected symmetrically by the GS lasing regime (dark red). Because of this symmetry, electron depletion then also leads to GS quenching. This symmetry is lost for increasing electron energy separation ΔE_e. Other parameters given in Tab. 6. ©(2015) IEEE. Reprinted, with permission, from [ROE14]

The influence of this asymmetry is shown in Fig. 4.4. The electron confinement ΔE_e was changed from a symmetric case (dark red area) to an increasingly asymmetric one. While for the symmetric case low electron occupations also lead to GS quenching, only the hole depletion side retains solitary ES lasing for the asymmetric energy structures (red, orange, yellow areas).

Next, reducing the depth of the confinement for both electrons and holes in Eq. (4.6) is equivalent to increased increasing the temperatures. Figure 4.5 shows, in accordance with the phenomenological explanation of carrier escape, that high temperatures lead to a broader solitary ES lasing regime, until GS lasing becomes altogether impossible (at 1030K, dark red area). Yet, the temperature differences required to achieve a significant change in the analytical boundaries of the ρ^{ES}-phase space lie well outside of the experimentally feasible tens of Kelvins. The self-heating

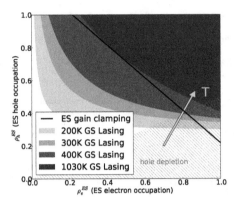

Figure 4.5: Two-state-lasing regimes as a function of excited state electron and hole occupation probabilities for different temperatures. The ES inversion clamping (black) intersects with the GS lasing regime (yellow to red) at high electron to hole ratios. With increasing temperature (darker colours) the two-state-lasing regime (overlap of ES gain clamping and GS lasing area) shrinks in size and vanishes for very high temperatures (dark red, 1030K for this set of parameters). Other parameters given in Tab. 6. ©(2015) IEEE. Reprinted, with permission, from [ROE14]

mechanism, proposed in the literature to explain GS-quenching [MAR03c, JI10], is therefore only a minor contributor in most cases. Every change in the electron-to-hole ratio has a larger impact than realistic temperature differences. Nonetheless self-heating can support the transition by widening the hole-depletion window for GS-quenching.

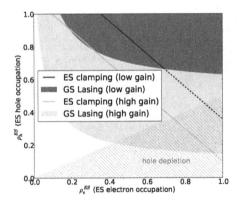

Figure 4.6: Gain dependence of GS quenching (possible on the dashed line). The low gain ($g_{GS} = 0.5g_{ES} = 0.07ps^{-1}$, red line) ES clamping is shifted to the right in comparison to the high gain scenario ($g_{GS} = 0.5g_{ES} = 0.25ps^{-1}$, orange line). The GS lasing regime is significantly larger for the high gain case (yellow area), as opposed to the low gain case (red area). Other parameters given in Tab. 6. ©(2015) IEEE. Reprinted, with permission, from [ROE14]

Lastly, the impact of the linear gain is shown in Fig. 4.6, where both the GS and ES gain were multiplied with the same factor. This is equivalent to a decrease in the optical losses κ (see Eq. (4.6)). Experimentally this corresponds to longer devices or higher QD densities. The change of size for the GS-lasing regime and solitary ES lasing regime is significant. While the low gain ($g_{GS} = 0.5g_{ES} = 0.07ps^{-1}$, red) exhibits solitary ES lasing for a wide range of ES occupations, the high gain ($g_{GS} = 0.5g_{ES} = 0.25ps^{-1}$, orange) virtually suppresses the entire GS-quenching window in the hole-depletion area. Yet despite a proposed gain decrease through homogeneous broadening [SUG05b], which has been later questioned by follow up studies [GIO12,

KOR13], the linear gain has to be treated as constant in our modelling approach. Choosing a gain of the correct magnitude is obviously of high significance, if one wants to facilitate two-state lasing, GS-quenching or solitary ES lasing. Yet, a current dependent, variable gain will not be considered from here on.

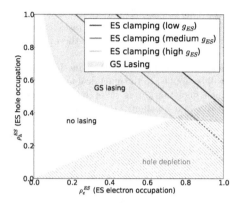

Figure 4.7: ES gain dependence of two-state lasing. When the ES gain g_{ES} is treated as an independent variable, the ES gain clamping (dark red, red, orange line) can be moved relative to the GS-lasing regime (yellow area). While the low ES gain ($g_{ES} = 0.115ps^{-1}$, dark red line) exhibits only two-state lasing, a higher ES gain ($g_{ES} = 0.23ps^{-1}$, red line and $g_{ES} = 0.46ps^{-1}$, orange line) enables solitary ES-lasing (dashed parts of the line). ©(2015) IEEE. Reprinted, with permission, from [ROE14]

When the ES gain is treated as a free parameter, instead of being always the double of the GS gain due to degeneracy, or if the ES and GS experience a different cavity, e.g. through spectral coating [ARS14], the ES gain clamping line can be easily moved relative to the GS-lasing regime. As seen in Fig. 4.7 high ES gain ($g_{ES} = 0.46ps^{-1}$, orange line) possesses a smaller two-state lasing regime and enables solitary ES lasing, whereas this is absent for low ES gain ($g_{ES} = 0.115ps^{-1}$, dark red line). The intermediate $g_{ES} = 0.23ps^{-1}$ is the reference gain assuming that $g_{ES} = 2g_{GS}$.

In summary, this semi-analytical approach allows to analyse the impact of parameter changes on the two-state lasing behaviour independent of the current-dependent occupation dynamics. Assuming realistic QD electronic structures, the window for GS-quenching is only present at low hole occupations. This is strong evidence that hole-depletion for high currents is the dominating effect leading to GS quenching. All other explanations as given in Sec. 3.3.1 play only a minor role.

The knowledge gained for choosing the right parameters will later be reflected in Sec. 4.3, where varying optical losses κ, ES gain g_{ES}, electronic confinements and temperatures T are investigated. These parametric studies are in good agreement with the analytic results that high κ, γ_{ES} and temperatures all favor solitary ES lasing, whereas strong confinement leads to a broader GS-lasing regime.

When these analytical boundaries are combined with numerically calculated, current dependent ES occupations $\rho^{ES}[J]$, the steady states of the system can be traced across the ρ^{ES}-plane. This will be done at the end of the following Sec. 4.2.

4.2. Numerical Simulation of GS Quenching

4.2.1. Modeling Approaches and Light-Current Characteristics

The numerical model introduced in Sec. 2.3 shall now be used to derive current-dependent steady states. The parameters used are the same as in Sec. 3.2 in Tab. 4. They were chosen based on previous works of Benjamin Lingnau [LIN14] and Ref. [LUE09, LUE11a, LUE12]. However, with the scattering rates as derived in Sec. 2.3.2, GS quenching is never reached. This can bee seen in the LI-curve of Fig. 3.3 on page 32.

Three different modelling approaches corresponding to the three mechanisms explained in the literature in Sec. 3.3.1 will be applied to the numerical model: Homogeneous broadening, self-heating and hole depletion. From these three the homogeneous broadening induced decrease of the gain constants g_{GS} and g_{ES} is the least physically sound. With active and inactive dots already included in the model and 'spectral holeburning' induced mechanics already approximated to a first order, any further change of gain parameters seems arbitrary.

Nevertheless, for completeness a short reproduction of the findings of Sugawara et al. [SUG05] shall be included. The gain of the GS and ES will be reduced with increasing overall intensity, accounting for a further broadening of the spectral line and an decreased overlap of QD ensemble and lasing wavelengths:

$$g_m = \frac{g_m^0}{1 + \varphi(|E^{GS}|^2 + |E^{ES}|^2)}, \qquad (4.8)$$

with $m \in \{GS, ES\}$ and g_m^0 being the gain at zero intensity. The suppression coefficient φ was set to $5 \cdot 10^4 [\text{V}/\text{nm}^2]^{-1}$ to yield GS quenching. Fig. 4.8 shows the resulting light-current characteristic. In agreement with expectations, the GS quenching is observed. Reducing the gain is the most efficient way of suppressing lasing activity, so the inclusion of an intensity-dependent gain will naturally lower the GS gain until only ES lasing is stable. However, as neither the underlying physical process nor the magnitude of the parameters can be satisfyingly deduced from a first principle, an increase of homogeneous broadening will not be included from here on.

Self-heating was included by increasing the out-scattering rates $S_{b,out}^{m,cap}$ and $S_{b,out}^{Rel}$ according to the detailed balance condition:

$$S_{out} = S_{in} e^{-\frac{\Delta E_k}{k_b T}}, \qquad (4.9)$$

where ΔE_k is the potential energy difference of the two energy levels involved in the scattering process. The temperature itself was modelled to increase linearly with pump current J, to account for the joule-heating of a device with constant voltage applied [LUE12]:

$$T = 300\text{K} + DJ, \qquad (4.10)$$

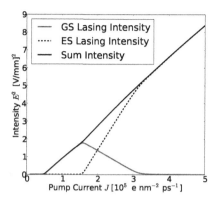

Figure 4.8: Simulated light-current characteristic for self-assembled InAs/GaAs QDs exhibiting GS quenching caused by homogeneous broadening. The GS and ES gains were modified to decrease with increasing current densities, to account for a strong increase of spectral width of lasing lines. The GS starts to decline after the onset of ES lasing, while the slope of the overall intensity stays roughly the same. Parameters as given in Tab. 4.

where the heating coefficient was chosen to be $D = 35[\mathrm{K}] \cdot 10^4[\mathrm{e/nm^2ps}]^{-1}$. The resulting GS-quenching light current characteristic is shown in Fig. 4.9. The temperature (green) rises linearly with pump current J (x-axis), while the GS intensity (red) starts to decline after the onset of ES lasing (blue). However, the temperature at which GS quenching is occurring is $T \simeq 450\mathrm{K}$, which is higher than what most devices are able to withstand.

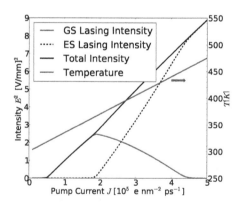

Figure 4.9: Simulated light-current characteristic for self-assembled InAs/GaAs QDs exhibiting GS quenching caused by self-heating. The temperature (green) was modelled to increase linearly with injection current J. The GS starts to decline after the onset of ES lasing, while the slope of the overall intensity stays roughly the same. Parameters as given in Tab. 4.

In favour of the self-heating hypothesis is the fact, that the carrier temperature could potentially be higher than the surrounding device temperature, as the pump current drives the system far away from thermal equilibrium. However, as already mentioned in Sec. 3.3.1, this thermal evolution should not be unique to InAs/GaAs QDs, but also be applicable to InP QDs, where GS quenching has not been observed. This leads to the conclusion that self-heating is not playing a major role in the appearance of solitary ES lasing and focuses the attention of this work on the third mechanism mentioned.

The electron-hole asymmetry leading to 'hole depletion' as first mentioned by Viktorov *et al.* [VIK05] shall now be reproduced. Subsequently, the scattering behaviour of our system was changed to achieve low hole densities by reducing the hole capture scattering rates. $S_{h,in}^{GS,cap}$ and $S_{h,in}^{ES,cap}$ were reduced to 5% of their microscopically calculated value. This is, of course, a great violation of the motivation behind calculating scattering rates microscopically and can only be justified in two ways: Either the real-world scattering rates still differ from the current microscopic model due to some processes not being accounted for, e.g. non-parabolic wavefunctions [SCH07f] or Coulomb interaction of carriers, or the energy levels used as initial preconditions for calculating the scattering dynamics are different from QDs that exhibit GS quenching. Also, Gioannini (2012) [GIO12] has shown a convincing alternative by introducing long transport times through an additional confinement structure. Including more hole states for the modelling can also help, as then by distributing the available charge carriers more evenly over the different states, the occupation of a single state is reduced.

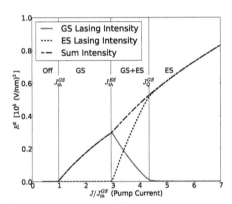

Figure 4.10: Light-current characteristic for self-assembled InAs/GaAs QDs exhibiting GS quenching caused by hole depletion. The GS intensity starts to decline after the onset of ES lasing, while the slope of the overall intensity stays roughly the same. The pump-current has been normalized to the GS lasing threshold J_{th}^{GS}. The ES lasing threshold J_{th}^{ES} and GS quenching current J_Q^{GS} are marked with vertical lines. Parameters as given in Tab. 4, with microscopically calculated scattering rates as in App. A.1 with hole capture reduced to 5%. [**Reference Parameter Set**] ©(2015) IEEE. Reprinted, with permission, from [ROE14]

Figure 4.10 shows the light-current characteristic for a hole-depletion induced GS quenching. This will serve as a reference simulation from here on to further study the hole-depletion mechanism in greater detail. As scattering rates and energy levels are highly material sensitive, the process of hole depletion is arguably the only remaining physical mechanism that sets different species of QDs apart. So, in accordance with the findings of the most recent literature [GIO12, KOR13, KOR13a], electron-hole asymmetry emerges as the major contributor for GS quenching. Therefore the rest of this chapter shall be devoted to further studying the hole-depletion induced GS-quenching mechanism.

4.2.2. Carrier Dynamics in GS Quenching

To further illustrate the driving mechanism between the transition of different lasing states, Fig. 4.11 plots the GS and ES occupations corresponding to Fig. 4.10 versus pump current. Here only the densities of the active subensemble are shown, as they are most relevant to the lasing behavior.

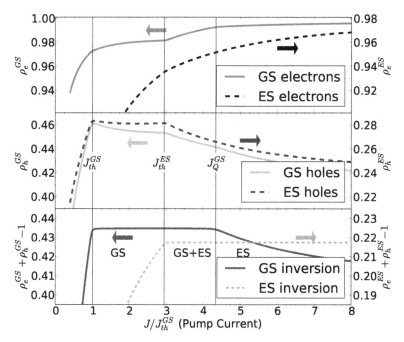

Figure 4.11: Electron (top panel) and hole (middle panel) occupation probabilities for ground (solid lines) and excited state (dashed lines) during GS quenching. Electron occupations are higher because of the higher energy spacing between QD and reservoir. Different lasing regimes are separated by vertical lines. During GS lasing the GS inversion (green solid line, bottom panel) is gain clamped until it quenches at J_Q^{GS}, the ES inversion (green dashed line, bottom) is clamped after the ES lasing threshold J_{th}^{ES}. Parameters as given in Tab. 4, with microscopically calculated scattering rates as in App. A.1 with hole capture reduced to 5%. [**Reference Parameter Set**] ©(2015) IEEE. Reprinted, with permission, from [ROE14]

Electron occupations (Fig. 4.10, top panel) are generally higher than hole occupations (Fig. 4.10, middle panel). This is caused by their higher electronic confinement of 50 meV, as opposed to only 20 meV for holes, resulting in smaller escape rates and higher equilibrium densities.

In the regime with no lasing ($J < J_{th}^{GS}$), all occupations increase with injection current J. This is caused by the overall increase of carriers and the resulting filling of states inside the reservoir, GS and ES. When GS lasing is reached at the GS threshold current $J = J_{th}^{GS}$, the GS inversion is clamped at $\rho_e^{GS} + \rho_h^{GS} = \kappa/g_{GS} + 1$ and will henceforth be constant (green solid line, bottom panel). Up until this point, the dynamics mirror a conventional laser reaching its threshold. Yet, as described in Sec. 3.2, there is no gain clamping of the ES by the GS occupations. Naturally, when extra charge carriers are added, many of them end up getting consumed by stimulated emission and increase the the GS lasing intensity above threshold (see Fig. 4.10, red line). But due to the incomplete gain clamping, carriers will also start to fill the ES (dashed lines).

This has already been mentioned as a requirement for two-state lasing; relaxation processes must not be too fast, as otherwise the GS and ES occupations would always be close to equilibrium and the ES would also be clamped [MAR03a]. The system needs to be allowed to retain extra charge carriers in the ES, while the GS is lasing. Two-state lasing is only possible because of the ES occupations increasing despite the GS inversion being already clamped.

However, with the increased complexity introduced through the non-excitonic nature of the numerical model, an additional detail starts to emerge: It is mainly the electron occupation probability ρ_e^{ES} that is rising, whereas hole occupations do not increase inside the GS lasing regime. This can only be interpreted in one way: The GS holes are clamping the ES holes, but the ES electron occupation ρ_e^{ES} is largely independent of the corresponding GS occupation ρ_e^{GS}. The extra holes added to the system are accumulating in the well w_h and the resulting increase in the scattering $S_{h,in}^{GS,cap}$ which *should* increase hole occupations is completely overcompensated by an increased carrier recombination through stimulated emission.

The incomplete gain clamping appears to be caused by the electrons in the system, while the hole occupations are much more closely tied to each other due to the smaller energy separation $\Delta E_h < \Delta E_e$. This is also in agreement with approaches applied previously in the literature: The analytical approximations made by Viktorov *et al.* [VIK05] and Korenev *et al.* [KOR13, KOR13a], who both combined the hole GS and ES into a common level, virtually achieve the same outcome. The numerical findings of this work also resemble the instant-equilibrium approach for hole-densities applied by Gioannini (2012) [GIO12].

When ES lasing starts at the ES threshold current $J = J_{th}^{ES}$, gain clamping occurs for it as well (green dashed line in Fig. 4.11, bottom panel). This would leave an excitonic model with no further degrees of freedom for the system, but in the non-excitonic picture of this work the fraction of electrons to holes can still change. Due to the higher scattering induced input, electron densities in the ES are still rising (dashed line, top panel). ES Gain clamping will then symmetrically lower ES hole occupation (dashed line, center panel) as $\rho_e^{ES} + \rho_h^{ES} = const$. Due to the equilibrating relaxation scattering processes, the GS is always bound to the ES and must follow the increasing electron fraction. Therefore, GS electrons are also rising while GS holes are decreasing (Fig. 4.11, middle and top panel, solid lines). This is exactly the behaviour that was assumed in the derivation of the analytical lasing boundaries of Sec. 4.1, namely that the GS occupations can be inferred from their ES counterparts.

With the increasing ES electron occupation ρ_e^{ES} rising and constantly forcing the GS electron occupation ρ_e^{GS} to follow it, ρ_e^{GS} soon reaches values above 0.99. Then, at the GS quenching threshold current $\left(J = J_Q^{GS}\right)$ electron occupations in the GS are virtually filled $\left(\rho_e^{GS} \simeq 1\right)$ and can no longer increase with rising J.

Simultaneously, GS holes are still depleting to follow the ES trend. At this stage, the quasi-equilibrium occupations of the GS will fall below the inversion needed to sustain lasing and thus force the GS to quench. The quenching is aided by the high Boltzmann-factor for the electrons, which enables the ES electrons to increase even further.

For high pump currents only solitary ES lasing remains. Holes deplete even further towards higher currents, but do not alter the lasing regime. Thus, the carrier dynamics shown in Fig. 4.11 and explained in the paragraphs above nicely exemplify the underlying mechanics of the inequality given in Eq. (4.6).

To conclude, there is not only 'hole depletion' but also a 'saturation of electrons' that leads to GS quenching. The GS has to compete with the excited state for carriers, and can only do so by holes, because GS electrons are saturated.

4.2.3. Comparison with Analytical Approach

The reference simulation with light-current characteristic in Fig. 4.10 on page 46 and density dynamics shown in Fig. 4.11 will now be visualized in a third way, by combining the numerical results with the analytical approximations derived in Sec. 4.1. The analytical lasing boundaries of Eq. (4.7) and Eq. (4.6) will be displayed with the numerically calculated, current dependent ES occupations $\rho^{ES}[J]$ of the reference simulation. The steady states of the system can then be traced across the ρ^{ES}-plane and the crossing of the analytical lasing regime boundaries should correspond to a change of lasing state in the numerics.

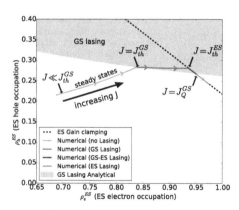

Figure 4.12: Analytically derived lasing regimes from Eq. (4.6) and numerically obtained steady state occupations ρ_e^{ES} and ρ_h^{GS} for increasing pump current J. Numerical results are colored according to the different lasing states. No lasing on the orange part of the line, red represents solitary GS lasing. When the excited state inversion is reached (black dashed line), two-state lasing happens on the dark red line and solitary ES lasing on the blue line. Parameters as in Tab. 4 on page 32, with microscopically calculated scattering rates as in App. A.1 with hole capture reduced to 5%. [**Reference Parameter Set**] ©(2015) IEEE. Reprinted, with permission, from [ROE14]

The steady states of the reference simulation in the ρ_e^{ES}-ρ_h^{ES} phase-space are shown in Fig. 4.12. Because the J-dimension is lost in this representation, the numerically derived line is color-coded according to the different lasing states.

The orange part of the line is below threshold, and as expected the numerical occupation probabilities lie outside of the analytical lasing regimes. Transition to

GS lasing (red line) is then observed as soon as the border of the GS lasing regime (yellow area) is crossed at $J = J_{th}^{GS}$. When the necessary excited state inversion is reached (black dashed line), the numerics will lead to two-state lasing (dark red line) at $J = J_{th}^{ES}$, and the system is hence forced to stay on the inversion given by the gain clamping. The fraction of electrons to holes is now the only degree of freedom left. Due to the low hole capture rates $S_{h,in}^{m,cap}$, the system approaches hole depletion and leaves the analytical GS lasing regime (yellow area) at $J = J_Q^{GS}$. This coincides with the transition to solitary ES lasing and GS quenching (cyan line) in the numerical simulation.

The agreement between analytical and numerical results in Fig. 4.12 is good, even though GS roll-over occurs for slightly lower hole occupations in the numerical model (barely visible in the plot). This can be explained by the direct capture processes from the surrounding carrier reservoir to the QD ground state, not taken into account in the analytical part. The direct capture processes slightly extend the GS lasing regime beyond the analytical approximations.

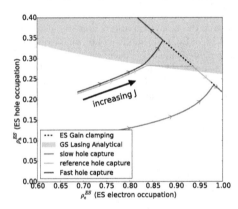

Figure 4.13: Analytical two-state lasing regime and numerical simulations of the steady states vs. excited state occupations for different hole capture scattering rates (different colours). Red are 0.5% hole capture rates, blue is our reference scattering and the pink line denotes a five times faster direct hole capture process (which is still only 25% of the microscopical scattering rates). Different regimes are crossed and lasing transitions differ accordingly. Parameters as in Tab. 4 on page 32. ©(2015) IEEE. Reprinted, with permission, from [ROE14]

The reason why hole capture rates had to be reduced to facilitate GS quenching will now be shortly highlighted, by changing the hole-capture rates and plotting the resulting J-dependant steady states in the ρ_e^{ES}-ρ_h^{ES} phase-space. As simulated in Sec. 3.2 for Fig. 3.3, with faster hole scattering rates GS quenching is absent as hole depletion is never reached. This is caused by the steady states moving to higher ρ_h^{ES} and into the stable two-state lasing regime. To illustrate this, Fig. 4.13 displays the results shown in Fig. 4.12 (cyan line) together with numerical simulations for slower (red line) and higher (pink) hole capture rates.

For the very slow hole capture process (red line), the hole occupations are suppressed so strongly that the GS lasing regime is never crossed. Consequently, the corresponding light-current characteristic then exhibits no GS lasing. This ES only lasing is similar to a high loss scenario, where light-current characteristics lack the GS transition as well. On the other hand, when hole capture rates are high (Fig. 4.13, pink line) - but still smaller than electron capture rates - the GS will never quench, as the electron-hole ratio shrinks with increasing currents. The two-state lasing state is stable for all pump currents and GS lasing intensity will even increase after the ES switch-on.

Looking at Fig. 4.13, one might ask the question for which magnitude of hole in-scattering the hole fraction starts to reduce after the onset of two-state lasing. Somewhere between the 'fast hole capture' (pink) and 'reference hole capture' (cyan) should lie a critical scattering rate, which leads to a constant electron-to-hole fraction. The derivation of this point could then possibly lead to some mathematical condition between hole and electron scattering rates, if hole depletion is to be reached. However, this was not achieved in the time of this work and it must therefore be left for future investigations.

Overall the analytical approximation has shown itself to be very robust, owing to the large difference between relaxation scattering and GS capture scattering magnitudes. This leads to a close tying of GS occupations to their ES counterparts, while simultaneously the ES is able to avoid gain clamping by being exposed to the much larger ES capture rates.

4.2.4. Turn-On Dynamics

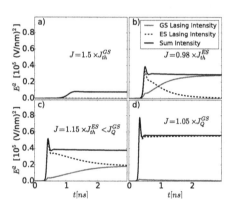

Figure 4.14: Electric field intensity turn on curves for ground and excited state lasing. Even for currents below the ES threshold (b), the ES is lasing during turn-on transient, but switches off again. During two-state lasing (c), GS is slower to converge as the ES levels fill up faster. For currents higher than the GS quenching threshold (d) GS lasing is shortly visible during relaxation oscillations. Parameters as given in Tab. 4, with microscopically calculated scattering rates as in App. A.1 with hole capture reduced to 5%. [**Reference Parameter Set**] ©(2015) IEEE. Reprinted, with permission, from [ROE14]

So far the time evolution of the electric field amplitude in the two optical modes (GS and ES) was not addressed. In this section the turn-on dynamics of the reference simulation shall be discussed for the different lasing regimes (see Fig. 4.10 for steady states). Quantum dot lasers exhibit strongly damped relaxation oscillations, as visible in Fig. 4.14 (a) for the GS turn-on inside the GS lasing regime.

However, during turn-on both optical transitions of GS and ES can be visible, even if the corresponding steady states are outside of the two-state lasing regime. As seen in Fig. 4.14 (b), ES lasing occurs temporarily for currents lower than the ES lasing threshold. Accordingly, GS lasing takes longer to increase. This is due to faster scattering into the ES levels, as well as due to the resulting slowly building up of the GS inversion clamping.

The turn-on delays of ES and GS can be analytically approximated from the effective carrier lifetime of the level and the modal gain [SOK12]. For our scattering rates and set of parameters the turn-on delay is equal for both ES and GS, so that both turn on synchronously. Note that the correlation of the relaxation oscillations, e.g. peaks coinciding in Fig. 4.14 for both lasing transitions, is caused by the fast relaxation from ES to GS. Overall, the shape and timing of the simulation is in good agreement with the experimental measurements of Ref. [DRZ10].

Inside the two-state-lasing regime at $J = 1.15 J_{th}^{ES}$, the ES is again starting to lase earlier (Fig. 4.14 (c)). Ground state lasing is also visible for currents greater than the quenching threshold (Fig. 4.14 (d)). During turn-on oscillations, ES and quantum well occupations will be higher than in the eventual steady state. This allows the GS to be filled above its threshold and the transition is visible for several ns. In an experimental setup this might be useful in finding a current range that is closest to achieving two-state lasing, e.g. if ES lasing can only be started via external injection or by introducing an additional feedback loop.

Strikingly, the overall intensity (black line (a)-(d)) converges significantly faster than the individual contributions of GS and ES. Due to the time constraints of this work the exact cause of this has not been found so far. However, from a purely dynamical standpoint it is clear to say, that within the high-dimensional phase-space of laser operation the system is highly damped transversal to the plane of $|E^{GS}|^2 + |E^{ES}|^2 = const$. But within this plane of constant overall-intensity the convergence is much slower and the real-part of the corresponding eigenvalues is supposedly closer to zero.

However, this explanation is lacking a physical mechanism explaining and quantifying the important damping factors. One might formulate the hypothesis, that charge carrier conservation forces the system to adjust its overall lasing-output to the incoming electron and hole flux on relatively short time scales. This would suggest that the damping of the overall relaxation is related to the traditional relaxation oscillation damping of single-state lasers, so that the usual calculus should be applicable. Simultaneously, the competition between modes is, arguably, linked to the difference of some 'effective gain'. Furthermore this behaviour is comparable with the turn-on dynamics of multi-mode lasers [DOK12] and is likely caused by a mechanism similar to that.

4.3. Lasing Regimes In Parameter Space

4.3.1. QD size and optical losses dependence

Inspired by the analytic results and the crucial role of the energy separations a systematic scan of the parameter space seems prudent. At first, the band structure will be continuously scaled. To do that, a linear scaling factor r is introduced. Energy spacing between levels are multiplied with r and new scattering rates are calculated for the resulting energy structure (see also App. A.2). Lower r corresponds to a more shallow energy structure, higher r for deeper QD levels (keeping the asymmetry). By looking at different steady states as a function of r, a qualitative picture for the two-state lasing behaviour of QD lasers with different sizes can be obtained.

Secondly, the optical losses κ are also varied, while keeping $g_{ES} = 2g_{GS} = 0.23ps^{-1}$ for the moment. For a better comparison with experimental findings, these optical losses can be converted into cavity lengths ℓ via the relation [LUE08]:

$$2\kappa = (2\kappa_{int} - \frac{\ln r_1 r_2}{2\ell})\frac{c}{\sqrt{\epsilon_{bg}}} \qquad (4.11)$$

Where ℓ is the cavity length, r_1, r_2 are the facet reflection coefficients ($r_1 = r_2 = 0.32$ for a GaAs-air surface), c is the speed of light and $\epsilon_{bg} = 14.2$ is the background permittivity in the cavity. Internal losses of $\kappa_{int} = 110m^{-1}$ in accordance with Ref. [LUE08] were used. This formula is only valid for the Fabry-Perot type edge-emitting lasers used in this work.

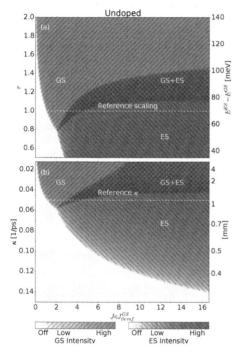

Figure 4.15: Lasing regime for GS (orange) and ES (blue) versus pump currents and confinement scaling factor r (a) and optical losses κ (b) obtained by numerical simulation. The two-state lasing is visible (cross hatched area). The reference lasing intensities for size $r = 1$ and $\kappa = 0.05ps^{-1}$ are seen in Fig. 4.10. GS quenching is only observed for some sizes r, others exhibit only ES lasing (shallow dots), no ES lasing (deep dots) or only a saturation of the GS intensity. Also visible is the decreased lasing threshold for lower losses. Parameters not varied are given in Tab. 4, with microscopically calculated scattering rates as in App. A.2 with hole capture reduced to 5%. ©(2015) IEEE. Reprinted, with permission, from [ROE14]

Fig. 4.15 (a) shows the lasing regimes for GS and ES for different QD confinement and pump current. The parameters corresponding to the reference light curve of Fig. 4.10 are marked by the white dashed line in the parameter-plots. White areas correspond to no active lasing, while orange areas exhibit GS lasing and blue areas ES lasing. The two lasing modes are also hatched differently, and their overlay is visible as a mixing of colours and the overlap of the hatching schemes.

As seen in Fig. 4.15 (a), for greater energy separation, e.g. higher r, no GS quenching will be observed and ES lasing might not even start. This is due to the high confinement leading to lower ES occupations, as the GS is energetically more favourable. On the other hand, smaller energy spacing suppresses GS lasing and enhances ES filling. As the ES has a higher gain, it will always dominate when occupations of ground and excited state levels are similar. Hence the lower regions of Fig. 4.15 (a) display pure ES lasing.

Figure 4.15 (b) shows the variation of optical losses κ. The white dashed line again denotes the reference of Fig. 4.10 for $\kappa = 0.05\text{ps}^{-1}$. As expected from the analytic results shown in Fig. 4.6, the lasing threshold increases for higher losses, e.g. shorter cavities, while the onset of two-state lasing decreases for higher losses. Above a certain loss value $\kappa \simeq 0.06\text{ps}^{-1}$, only ES lasing can be observed, as the GS gain gets too weak to counter the optical losses. This corresponds to the parts of Fig. 4.6, where the GS lasing regime recedes further for lower gain, until only solitary ES lasing is observed. Contrastingly, for low losses (long cavities), the overlap between ES and GS lasing regime is so large, that the electron-hole ratio never surpasses the critical value necessary for GS quenching. Followingly the upper regions of Fig. 4.15 (b) exhibit stable two-state lasing.

As discussed in Sec. 3.3.1, this is in agreement with the experimental findings of Ref. [MAR03a]. Long cavities have lower losses κ (see second y-axis in Fig. 4.15 (b)). They found a critical length ℓ, below which only ES lasing was present, intermediate lengths with two-state lasing and an increasing threshold current for the ES lasing for larger devices. This has also been independently confirmed in Ref. [CAO09] and Ref. [LEE11c] and the need of short cavities is also mentioned in Ref. [VIK07a]. This is nicely reproduced by the parameter studies of Fig. 4.15. Additionally, Ref. [MAX13] shows the GS and ES threshold currents versus cavity lengths as measured by Maximov et al. and this also exhibits a good agreement with the numerical findings of this work. Higher cavity lengths favour GS lasing and below a certain critical length, only ES lasing is present.

4.3.2. Influence of Doping

Doping has been shown to influence two-state-lasing behaviour [MAX13] and lasing thresholds [TON06]. This will now be investigated further with the numerical model and be compared to the undoped case.

Dopings within this work have been simulated with 10 extra charge carriers per QD. Charge conservation is now maintained with an offset, accounting for the extra electrons or holes added by doping [LUE10, KOR10]. Increased intrinsic losses R_{loss}^{W} for the QW or photons κ_{int} were not used, even though they could be introduced by the higher defect rate in doped materials. Yet the focus shall be set on the charge carrier dynamics, and not be complicated by changing more than one parameter at a time.

When comparing the lasing regimes of Fig. 4.16 to the undoped Fig. 4.15, n-doping increases the lasing threshold. This is in accordance with previous theoretical results [TON06] and can be explained by the different hole and electron dynamics. As seen in Fig. 4.11, electron states are always fuller than their hole counterpart. So

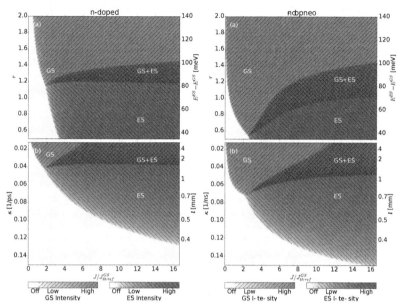

Figure 4.16: Lasing regime for GS (orange) and ES (blue) versus pump currents J and confinement scaling factor r (a) and optical losses κ (b) for n-doped QDs. The two-state lasing is visible (cross hatched). The n-doped QDs show a higher lasing threshold as compared to undoped (Fig. 4.15) and a smaller two-state lasing regime. Solitary ES lasing is more common, as the high electron to hole ratio needed for GS suppression is already intrinsically present. Parameters as given in Tab. 4, with microscopically calculated scattering rates as in App. A.2 with hole capture reduced to 5%. ©(2015) IEEE. Reprinted, with permission, from [ROE14]

Figure 4.17: Lasing regime for GS (orange) and ES (blue) versus pump currents J and confinement scaling factor r (a) and optical losses κ (b) for p-doped QDs. The two-state lasing is visible (cross hatched). The p-doped QDs exhibit a lower lasing threshold as compared to undoped QDs (Fig. 4.15) and a broader two-state lasing regime. GS lasing is enhanced and the ES lasing threshold is higher. Parameters not varied here are given in Tab. 4, with microscopically calculated scattering rates as in App. A.2 with hole capture reduced to 5%. ©(2015) IEEE. Reprinted, with permission, from [ROE14]

naturally, adding more electrons does not significantly increase occupations in the QD, but mostly in the well w_e. This leads to higher non-radiative loss processes, which always also remove a hole from the system. Therefore, n-doping increases losses without aiding lasing, which leads to an overall increase in the lasing threshold.

Somewhat counterintuitively, GS quenching is observed for a smaller range of parameters. On first thought, one could suspect that the additional electrons introduced by n-doping are aiding the hole-depletion process. So a broader range of GS quenching could be expected. That this is not the case in Fig. 4.16 can be explained by the broader ES lasing regime. In general, doping becomes less important for high pump currents J, so that n-doped and undoped will converge to similar electron-hole ratios far above lasing threshold. This is reflected by the fact that the high -current right-hand sides of the n-doped Fig. 4.16 and the undoped Fig. 4.15 look the same. On the other hand, n-doping significantly influences the low current dynamics. The n-doped samples already start with more electrons than holes and will then start to lase directly in the ES, with no intermediate two-state lasing. GS quenching is therefore not observed, because the GS already starts suppressed for a broader parameter range.

Oppositely, adding additional holes via p-doping (Fig. 4.17) leads to a smaller lasing threshold. Most holes that are intrinsically present will relax into the QD GS and aid the onset of GS emission. As opposed to electron occupations, hole occupations are far from saturation, so adding additional holes fills the GS faster.

P-doping also leads to higher GS output power and broader GS lasing regime. Opposite to the effects of n-doping, initial GS lasing and subsequent two-state lasing is observed for a greater parameter range. This is once again caused by the doping-carriers dominating the low-current region, where additional holes facilitate GS lasing. Subsequently, the steady state solutions converge to the undoped-cases when injected carriers start to outweigh the doping carriers for high injection currents J. Following, GS quenching is observed for a broad parameter range. P-doping therefore enhances GS quenching.

The reduction of the GS lasing threshold by p-doping has also been theoretically predicted by Ref. [JIN08] for some parameter sets. On the contrary, the experiments of Ref. [ALE07] and [MIK05] show an increase of the lasing threshold for p-doping, but this is attributed to the increased optical losses, so that the results of the numerical simulation are confirmed by real world QD behaviour.

4.3.3. Temperature and ES gain dependence

Furthermore, the background temperature T and ES gain g_{ES} are also parametrically studied. Here the degeneracy caused restriction of $g_{ES} = 2g_{GS}$ used so far is lifted and the ES gain is treated as an independent parameter. On a microscopic level this can compensate the effects of different electric dipole moments for the two possible optical transitions and, on the other hand, can also model the scenario of different mirror reflectivities for the GS and ES wavelengths.

The temperature enters the numerical simulations via the detailed balance condition, modifying the Fermi-function and hence the difference of in- and out-scattering. A higher temperature will lead to a broader Fermi-distribution and therefore equalise the GS and ES occupations, while a lower temperature leads to a concentration of carriers in the GS. Also, there is no self-heating included in these simulations. Temperature changes of more than 50 K would also significantly change the scattering behaviour of the in-scattering rates, The ES gain variation directly influences the ES lasing threshold, but has otherwise no direct effect on the system.

Figure 4.18 (a) shows the lasing intensities for GS (orange) and ES (blue) versus pump current and background temperature for undoped QDs. In contrast to the cavity length and QD size plots, the transition from two-state lasing to solitary ES lasing is less pronounced in the temperature plots. Especially in the lower tempera-

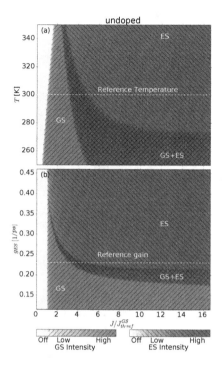

Figure 4.18: Lasing regime for GS (orange) and ES (blue) versus pump currents J and temperature T (upper panel) and ES gain g_{ES}. The ES gain was treated as an independent parameter, so that at $g_{ES} = 0.115ps^{-1}$ (lowest limit of (b)) it is equal to the GS gain g_{GS}. The reference lasing intensities for size $T = 300$K and $g_{ES} = 0.23ps^{-1}$ are seen in Fig. 4.10. GS quenching is only observed for some temperatures and gains, others exhibit only ES lasing (high ES gain), no ES lasing (low ES gain) or only a saturation of the GS intensity (lower temperatures). Also visible is the decreased lasing threshold for lower temperatures. Parameters not varied here are given in Tab. 4, with microscopically calculated scattering rates as in App. A.2 with hole capture reduced to 5%.

ture regions ($\sim 270K$) GS lasing is sustained over a broad current range. Therefore, changing the temperature significantly changes the extent of the two-state lasing regime, with a faster GS quenching at higher temperatures. Low temperature delays the onset of ES lasing as well and lowers GS lasing thresholds.

This is in good agreement with the experimental results of Maximov *et al.*. Fig. 4 of Ref. [MAX13] shows their experimental measurements of the threshold current densities for the GS (circles) and ES (squares). They also measured p-doped QDs and its effect on two-state lasing.

The ES gain dependence of Fig. 4.18 (b) is as expected: Higher g_{ES} enhances the ES lasing intensity and reduces ES threshold currents, while lower g_{ES} delays the onset of two-state lasing. Note, that on the lower border $g_{ES} = g_{GS} = 0.115ps^{-1}$ and that ES lasing is absent. Furthermore, it is visible that the GS lasing threshold is independent of ES gain - as the border between GS lasing regime (red) and no lasing (white) is vertical in (b).

So as done in the previous section, a p-doped and an n-doped QD ensemble is also simulated for different temperatures and ES gains. Fig. 4.19 shows the p-doped parameter plots. GS lasing is once again enhanced by the p-doping of QDs, as the holes are the rare species of carriers. The additional holes in the system delay hole-depletion, so that a wide array of parameters start to lase on the GS and then switch to the ES. ES lasing thresholds are greatly increased, which is in good qualitative agreement with Fig. 4 of Ref. [MAX13], where the temperature dependence of a p-doped sample was also compared to an undoped QD sample.

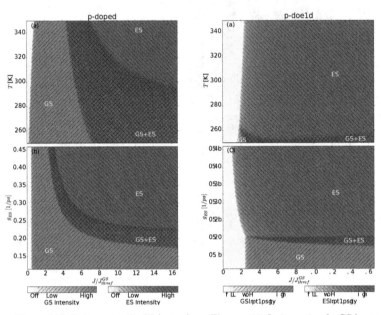

Figure 4.19: Lasing regime for GS (orange) and ES (blue) versus pump currents J and Temperature T (a) and ES gain g_{ES} (b) for p-doped QDs. In comparison to the undoped case (see Fig. 4 of Ref. [MAX13]), ES lasing is weakened and GS quenching happens for a broader set of parameters. Also visible is the overall decreased lasing threshold.

Figure 4.20: Lasing regime for GS (orange) and ES (blue) versus pump currents J and Temperature T (a) and ES gain g_{ES} (b) for n-doped QDs. In comparison to the undoped case (see Fig. 4 of Ref. [MAX13]), ES lasing is enhanced and GS quenching happens for fewer parameters. Also visible is the decreased lasing threshold for higher ES gains, if and only if the ES is the first to lase.

Finally, n-doping increases thresholds and enhances ES lasing as shown in Fig. 4.20. The independence of the GS lasing threshold from the ES gain g_{ES} can also be seen in (b), where the overall lasing thresholds is only increased if and only if the ES is the first to lase. As soon as the ES lasing regime (blue) borders the no-lasing regime (white), the threshold currents starts to decrease with increasing ES gain. There also appears to be a region in temperature space, where a reappearance of the GS can be observed. The initial n-doping must be suppressing the hole fraction to such a low value, that it actually recovers for higher values. This would, in any experiment, probably be completely washed out by the high fraction of defects introduced through doping and not be observable.

To conclude these numerical parameter studies, it has become clear that GS quenching is a transition phenomenon in parameter space. GS quenching is the specially tuned case of parameters that lie inbetween the regions of purely ES lasing and purely GS lasing devices. A stable two-state lasing over a broad current range can be achieved by the right choice of parameters and might be of interest for anybody who wishes to fabricate two-state lasing devices. As opposed to QD size and ES gain, the cavity length and operation temperature can be changed during experimental operation and the numerical results presented in this section are in good agreement with the published experimental data.

5. Modulation Response

This chapter covers the modulation response of two-state quantum dot lasers, a quantity that can be easily measured and that is of crucial importance for communication applications [GRE13, LIN12, GIO11].

5.1. Data Transmission with Semiconductor Lasers

Semiconductor lasers are widely and commmonly used in optical data communication networks [GIO11, LIN12, GRE13], among other things as amplifiers [MEU09, SCH12e] or as mode-locked devices [ARS13]. As part of the ever expanding optical fiber networks, they are on the verge of becoming the backbone of modern information technology. Not only are all major parts of the long-distance connections, all submarine cables, no longer based on electronic transmission through copper lines, but optical fibers nowadays are also being used for internet access in private homes; in what the industry has termed 'fiber-to-the-home' connections [BON11].

Lasers based on self-assembled QDs possess a variety of advantages over other semiconductor systems and make them especially suited for these kinds of applications [BIM08]. They can emit at wavelengths of 1.3 μm and 1.5 μm, both of which are important as they represent minimum-loss cases for the currently installed optical fiber systems. With low threshold currents, high efficiency and long life times, QD lasers are among the most energy efficient lasers that are available. Additionally, temperature stability is very high caused by the discrete set of electronic states confined inside the QD box-like potential.

However, the most important factor is, of course, the maximum data transmission rate that can be achieved. In that regard QDs, at least in a Fabry-Perot-type device with no additional fabrication, do not reach as high of a modulation response as was theoretically predicted with simplified models in the 1980's [DIN76, ARA82, ASA86] as the finite time scattering processes limit the maximum carrier modulation that can be induced via varying the injection current [BIM08]. Current devices are able to reach error-free rates of up to 15 GHz [ARS14] as also theoretically described in [LIN12, LUE10a, LUE12].

Future improvements include vertical-cavity surface-emitting lasers, where a set of distributed Bragg-reflectors increases the reflectivity of the laser-cavity mirrors to above 99.9%. Here modulation rates of over 40 GBit/s [HOF11](Zitat) were achieved. Yet, out-coupling efficiencies and lasing intensities are low for such devices, as the high reflectivity leads to the concentration of lasing intensity only inside the active zone. Also actively being developed are electro-optical modulators, which offer a different approach. No longer relying on the carrier dynamics of the QD, the electriec field intensity is modulated directly, e.g. through voltage dependent absorption via the quantum-confined stark-effect, and even higher frequencies might be reached [WEG14].

Two-state lasing has so far not been in the focus of modulation-response related work. Only the modulation response of the ES has been compared to the GS [GIO06, VES07, ARS14], but no thorough current-dependent analysis has been published so far. This will be done by numerical simulation in the following sections.

5.2. Modelling of Modulation

When talking about modulation response, one has to make a distinction between a small-signal analysis and the larger modulation used for actually transmitting digital data. While the first is closely linked to perturbation induced relaxation oscillations and mimics a linear stability analysis, the latter includes more complex dynamics. In large signal analysis hysteretic effects can destroy the signal transmission at high frequency [LUE10a].

The small-signal modulation response of a QD device can be easily calculated with the numerical model. In experiments devices are fed a periodically modulated signal in the injection current. This can be included in the model by making the injection current J time dependent

$$J(t) = J^0 + \Delta J \sin\left(2\pi f t\right), \tag{5.1}$$

with a base current J^0, modulation amplitude ΔJ and modulation frequency f.

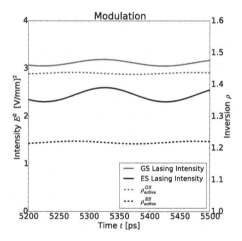

Figure 5.1: GS intensity (red) and active QD inversions $\rho_{act}^m = \rho_e^m + \rho_h^m$ for GS (green) and ES (blue) versus time. The injection current was modulated via $J(t) = J^0 + \Delta J \sin\left(2\pi f\right)$ and the resulting modulation of lasing field and carriers is shown for $2\pi f = 5$ GHz, $J^0 = 4 \times 10^{-5}\mathrm{enm}^{-2}\mathrm{ps}^{-1}$ and $\Delta J = 0.5 \times 10^{-6}\mathrm{enm}^{-2}\mathrm{ps}^{-1}$. This is a large modulation used to visualize the effect of a periodically varied injection current. Parameters as in table 4.

Figure 5.1 shows the resulting modulation of GS and ES electric field for $2\pi f = 0.5$ GHz, $J^0 = 4 \times 10^{-5}\mathrm{enm}^{-2}\mathrm{ps}^{-1}$ and $\Delta J = 5 \times 10^{-6}\mathrm{enm}^{-2}\mathrm{ps}^{-1}$. A periodic intensity fluctuation of the GS mode (red line) is achieved, while the relative fluctuation in carrier density is smaller. Even though the median ES intensity (blue line) is smaller, it is more strongly modulated by the injection current. This difference in the response is caused, in parts, by the faster capture channel for the ES carriers and the resulting stronger modulation of the ES inversion $\rho_{act}^{ES} = \rho_e^{ES} + \rho_h^{ES}$.

For visualization of this, Fig 5.2 shows the modulation response for identical parameters, but changed scattering rates. Here, the relaxation process S_b^{rel} was set to zero. To compensate for the loss of GS carrier input, GS capture $S_b^{GS,cap}$ was multiplied with a factor of 6. Detailed balance relations, however, were constantly

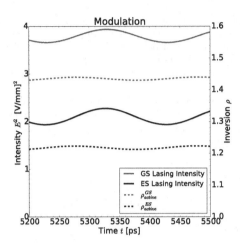

Figure 5.2: GS intensity (red) and active QD inversions $\rho_{act}^m = \rho_e^m + \rho_h^m$ for GS (green) and ES (blue) versus time for identical parameters as in Figure 5.1, but changed scattering rates. The relaxation scattering, meaning the direct exchange of carriers between ES and GS, was turned off and GS capture was speed up by a factor of six in this simulation. This illustrates the influence of the 'directness' of the carrier capture process on the resulting lasing intensity modulation. With GS capture now roughly twice as fast as ES capture, both resulting modulation amplitude are comparable in size.

maintained. As a result, the GS response is on the same order of magnitude as the ES response, illustrating that a direct, fast channel to the carrier reservoir is preferable to increase the amplitude of the modulated lasing intensity.

However, the fact that the GS capture has to be almost twice as fast as ES-capture also highlights that ES-amplitude is intrinsically stronger. While the scattering rates S determine how fast a given state equilibrates with the thermal distribution in the other carrier states, the equilibrium occupation ρ_{eq} itself also changes with injection current. As the Fermi distribution is less prone to perturbations for energies far from the Fermi-energy E_f, steady-state ES-carrier occupations have a stronger current-dependence. When the injection current is varied, this leads to higher carrier fluctuations than the GS, even if scattering time scales are the same.

The current-modulation amplitude ΔJ used here was rather large, to illustrate the effect of a sinusoidal signal on the injection current. In the following section, where frequency dependent amplitude responses will be calculated, a smaller ΔJ will be used, which is more in line with a small signal analysis.

5.3. Modulation Response Curves

The modulation response curve visualises the ability of the laser system to transfer a signal of the injection current into light intensity. As in the previous section the pump current was varied with $J(t) = J^0 + \Delta J \sin(2\pi f t)$, and the resulting intensity response $\Delta I^m = \Delta \|E^m\|^2$ with $m \in \{GS, ES\}$ is evaluated as a function of modulation frequency f. An example for parameters as in Tab. 4 and $J = 1 \times 10^{-5} enm^{-2} ps^{-1}$ is shown in Fig. 5.3.

The modulation response is normalized with respect to the intensity response at low frequency ($f = 20$ MHz in this case). Which corresponds to the system relax-

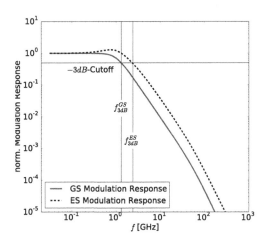

Figure 5.3: Normalised modulation response versus frequency f obtained by numerical simulation. The GS (red) and ES (blue) amplitude responses were normalized with respect to their low-frequency response. Where the relative strength of the light-intensity modulation drops below ~ 0.5 lies the 3dB-cut-off frequency f_{3dB}^m. Modulation responses can exhibit a resonance-feature, like the ES response in this picture, or simply diminish for high frequencies, like the GS here. Parameters as in Tab. 4, $J^0 = 1 \times 10^{-5}\mathrm{enm^{-2}ps^{-1}}$ and $\Delta J = 2 \times 10^{-7}\mathrm{enm^{-2}ps^{-1}}$.

ing to a steady state during each part of the modulation cycle; $I^m|_{J=J^0} - \Delta I^m = I^m|_{J=J^0-\Delta J}$. Contrarily, in the limit of very high frequencies ($f > 100GHz$ in Fig. 5.3) the system is far from reaching a steady state during one period of the modulation. Fluctuations are too fast for carrier populations to rise or fall. The system experiences only the median current J^0 and modulation is very weak. Therefore, the modulation strength is constant for the range of low frequencies and drops towards high frequencies.

For the right choice of parameters the lasing system can exhibit relaxation oscillations. These are tied to the eigenvalues of the system of differential equations. As the small-signal modulation disturbs the system only slightly, a linearisation of the system around the stable fix point for $J = J^0$ is a good approximation. Therefore, the damping and frequency of the relaxation oscillations are closely tied to the modulation response [LIN12].

In Fig. 5.3 this is visible for the ES in the shape of a resonance feature. For frequencies of about 1 GHz, the modulation response is exhibiting a maximum. By resonant excitation of relaxation oscillations the system's response is therefore greatly enhanced. Furthermore, this feature enhances the performance for faster modulation. The most important figure of merit in that regard is the '3dB-cut-off frequency' f_{3dB}^m. It is defined as the frequency where the relative strength of the light-modulation drops by 3 dB, which translates to a factor of $-3\mathrm{dB} = 10^{-0.3} \simeq 0.5$. As a first estimate, this corresponds to the maximum frequency where almost error-free data transmission is still possible. Because the GS exhibits no relaxation oscillations and followingly lacks the resonance feature of the ES, the GS cut-off-frequency f_{3dB}^{GS} is lower than the ES cutoff f_{3dB}^{ES} (see Fig. 5.3, vertical lines).

Modulation response curves are often measured experimentally, as the amplitude of light intensity modulations can be easily obtained [GRE13, LIN12, GIO11]. They are used in characterising a grown semiconductor laser sample and many publications, PhD theses and work groups are dealing with the right growth parameters for optimizing the laser performance, which usually means they try to reach as high values of the 3dB-cut-off frequency f_{3dB} as possible.

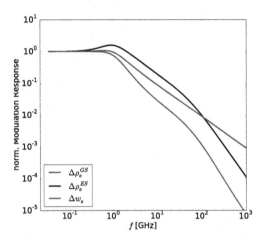

Figure 5.4: Normalised modulation response of w_e, ρ_e^{GS} and ρ_e^{ES} versus frequency obtained by numerical simulation. The excited state electrons (blue) exhibit a resonance feature, while both QW electrons and GS electrons exhibit none. Parameters as in Tab. 4, $J^0 = 1 \times 10^{-5} enm^{-2}ps^{-1}$ and $\Delta J = 2 \times 10^{-7} enm^{-2}ps^{-1}$.

For further study, Fig. 5.4 plots the normalised modulation response of QW electrons w_e, GS electrons ρ_e^{GS} and ES electrons ρ_e^{ES} versus frequency f. Defined analogously to intensity fluctuation, these carrier modulation responses are usually not readily available in experiments, but can be easily obtained in numerical simulations. The resonance feature of ES electrons matches the resonance of ES light intensity in Fig. 5.3. This is clear evidence of its relaxation-oscillation caused origin, as these oscillations stem from the periodic energy transfer between electron-hole pairs and photons. Furthermore, no resonance is visible for the two other electronic state occupations, which in turn do also exhibit strongly damped relaxation oscillations.

While the GS carrier dynamics are more complex, the shape of the w_e-modulation response is simply caused by the interplay of carrier-decay R_{loss}^w and the periodic injection current [LIN12], with the ES being additionally dominated by its resonance features. The analytic form for these response curves, which can be seen as an approximation of first order for all state variables, shall shortly be derived here.

Let $z \in \mathbb{C}$ denote a state variable and furthermore let its time evolution be described by three terms: First, a linear decay with time constant T; Second, an internal oscillation with frequency ω_{int}, e.g. modelling the periodic energy exchange with photons; Third, a time-dependant injection current $J(t) = J_0 + \Delta J \times e^{i\omega t}$ with external modulation frequency ω as a source. Thus, the differential equation for z is a driven harmonic oscillator:

$$\dot{z} = -\frac{z}{T} + i\omega_{int}z + J_0 + \Delta J \times e^{i\omega t}. \tag{5.2}$$

The steady state z_0 without modulation can then be derived via:

$$0 = -\frac{z_0}{T} + i\omega_{int}z_0 + J_0$$

$$z_0 = \frac{J_0 T}{1 - i\omega_{int}T} \tag{5.3}$$

The time-dependant dynamics can be written as the steady state solution z_0 and a small time-dependant perturbation $\delta z(t)$. Inserting $z(t) = z_0 + \delta z(t)$ into Eq. (5.2) yields:

$$(z_0 \dot{+} \delta z) = -\frac{z_0 + \delta z}{T} + i\omega_{int}(z_0 + \delta z) + J_0 + \Delta J e^{i\omega t}$$

$$\dot{z}_0 + \delta \dot{z} = -\frac{z_0}{T} + i\omega_{int}z_0 + J_0 - \frac{\delta z}{T} + i\omega_{int}\delta z + \Delta J e^{i\omega t}$$

$$\delta \dot{z} = -\frac{\delta z}{T} + i\omega_{int}\delta z + \Delta J e^{i\omega t}. \tag{5.4}$$

With the simple ansatz that z oscillates with the same frequency as the injection current, $\delta z = \Delta z e^{i\omega t}$, Eq. (5.4) becomes:

$$i\omega \Delta z e^{i\omega t} = -\frac{\Delta z}{T}e^{i\omega t} + i\omega_{int}\Delta z e^{i\omega t} + \Delta J e^{i\omega t}, \tag{5.5}$$

from which, after a short reshuffling, the complex modulation response $\Delta z/\Delta J$ can be extracted:

$$\frac{\Delta z}{\Delta J} = \frac{T}{1 + i(\omega - \omega_{int})T}, \tag{5.6}$$

and taking the absolute value yields:

$$|\frac{\Delta z}{\Delta J}| = \frac{T}{\sqrt{1 + (\omega - \omega_{int})^2 T^2}}, \tag{5.7}$$

which is the approximate shape for the modulation response curves. The drop-off towards high currents is furthermore given by:

$$\lim_{\omega \to +\infty}|\frac{\Delta z}{\Delta J}| = \frac{T}{\omega}, \tag{5.8}$$

which, in a double logarithmic plot, returns a straight line of slope -1.

Figure 5.5 shows w_e, ρ_e^{GS} and ρ_e^{ES} modulation response curves together with fits obtained via Eq. (5.7). The agreement for QW densities is good, which is also what can be expected as the reservoir carriers are directly modulated and their time

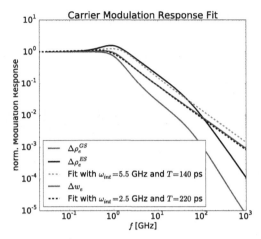

Figure 5.5: Normalised modulation response of w_e, ρ_e^{GS} and ρ_E^{ES} versus frequency obtained by numerical simulation, and analytical fits with Eq. (5.7). The fit for the directly modulated w_e-reservoir is shown by black dotted line, while the ES carriers' response is shown in light blue. No fit for the GS response could be obtained. Parameters for the numerical simulation as in Tab. 4, $J^0 = 1 \times 10^{-5} \mathrm{enm}^{-2}\mathrm{ps}^{-1}$ and $\Delta J = 2 \times 10^{-7} \mathrm{enm}^{-2}\mathrm{ps}^{-1}$; Fit parameters given in legend.

evolution most closely resembles the differential equation for z given in Eq. (5.2). However, the w_e decay rate of the numerical model in Eq. (2.68) is proportional to not only w_e, but also w_h. Which on a side note highlights, that the entirety of non-excitonic dynamics was not taken into account in the derivation of the fit of Eq. (5.7).

Despite this limitation, ES occupation-probability modulation $\Delta\rho_e^{ES}$ can be approximately reproduced with this fit. The resonance feature is weaker, yet overall agreement is good up to frequencies of $f \simeq 50$ GHz. At that point ES response is decaying even faster towards higher frequencies and is deviating from the previously predicted slope of -1 for high injection currents. As a result of previous works in the group of the author, it has been shown that this is linked towards the breakdown of carrier transport via scattering. Modulations do no longer propagate fast enough towards the confined QD states, but are kept within the QW-reservoir level. This can also be seen by the fact, that the w_e-response does not exhibit this scattering-related modulation break-down, as it is directly modulated via J.

Lastly, the ρ_h^{GS} modulation response of Fig. 5.5 was not reproduced. The dynamics here are clearly more complex than the simplistic approximations used for the fit. This is most probably related to the scattering dynamics, especially the strong dependence on ES occupations through the cascade scattering process QW-ES-GS. However, the modulation response drop-off at high frequencies can also be observed for the GS.

5.4. Cut-off-Frequencies and Two-State Lasing

After obtaining individual modulations and evaluating the response amplitudes, modulation response curves were introduced in the previous section. The 3dB-cut-off-frequency was introduced as the key figure of merit, often used in experiments to describe the maximum data transmission capacity of the device. Followingly, in this section modulation response curves will be numerically calculated and their cut-off-frequencies be obtained as a function of pump current J, among other parameters. Of special interest for this work is the interaction of modulation response with two-state lasing and GS quenching. The literature has not covered this topic extensively, and experimental verification of the predictions made here is therefore still lacking.

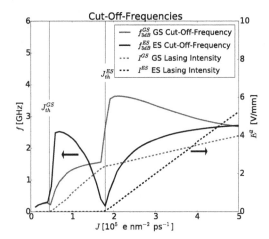

Figure 5.6: 3dB cut-off-frequencies for GS (red) and ES (blue) versus pump current obtained by numerical simulation. Light-current characteristic of the simulation shown in dashed lines. Parameters for the numerical simulation as in Tab. 4, $\Delta J = 2 \times 10^{-7} enm^{-2}ps^{-1}$ and nonlinear scattering rates as in App. A.1.

Figure 5.6 shows the 3dB cut-off-frequencies for GS f_{3dB}^{GS} and ES f_{3dB}^{ES} as a function of pump current J. Also plotted as a reference is the light-current characteristic with dashed lines. There is clearly a connection between lasing thresholds and modulation response, as cut-off-frequencies obviously react when approaching lasing states. The individual details will be discussed in the following paragraphs.

The sub-threshold ES (blue solid line) cut-off-frequency reaches relatively high values with $f_{3dB}^{ES} > 2GHz$, yet as the ES is not lasing at this injection current, this is not useful for data transmission. As the ES lasing threshold is approached, modulation response is slowed down and reaches a minimum at $J = J_{th}^{ES}$. This can be partly tied to the relaxation oscillations slowing down at the threshold and partly attributed to the long effective carrier lifetime at threshold. The cut-off-frequency increases for injection currents and then saturates in agreement with previous works of QDs with only one confined state [GIO11, LUE12].

The GS dynamics in Fig. 5.6 exhibit a similar shape. Below GS threshold, GS modulation response is slow, reaches a minimum for $J = J_{th}^{GS}$ and increases non-linear afterwards. However, the most striking feature is an almost vertical increase in the cut-off-frequency by a factor of two, when two-state lasing starts at $J = J_{th}^{ES}$. This is a phenomenon for which experimental verification is not available as of yet. Hence, from here on 'GS-modulation enhancement' will denote this sudden increase in the cut-off-frequencies for GS f_{3dB}^{GS} at the ES threshold. For high currents, the GS-modulation enhancement diminishes again and the GS cut-off-frequency approaches a static value (see also Fig. 5.12 for high-current behaviour).

5.5. Ground State Modulation Enhancement

Investigating this ES-lasing-induced GS-modulation enhancement is not finished and further research is necessary. Investigations are made difficult by the fact, that relaxation oscillations are almost nowhere present in the parameter ranges studied so far, as is generally the case for QD-based devices [BIM08a]. Numerically evaluating their damping and frequency, however, could have given a first clue towards the nature of this modulation enhancement. A drastic increase of GS relaxation oscillation frequency at the ES threshold current could explain the abrupt enhancement of the 3dB-cut-off-frequency.

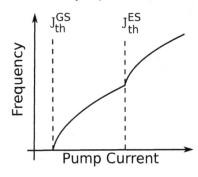

Figure 5.7: Relaxation oscillation frequency versus pump current as calculated by Abuusa *et al.* in [ABU13], schematically redrawn here. The drastic increase in the frequency coincides with the start of two-state lasing. Only a common frequency for both ES and GS was calculated.

However, an analytical approximation for the frequency of relaxation oscillations in two-state lasing lasers was published by Abuusa *et al.* in 2013 [ABU13]. They predict an increase of relaxation oscillation frequency for currents above the ES lasing threshold. The relevant figure from their work is schmetically redrawn here in Fig. 5.7.

However, it is questionable whether their analytical derivation is valid for the QDs simulated for this thesis. They only calculated a common relaxation oscillation frequency for both GS and ES, as they found no dynamical difference between both modes. This is in clear contrast with the rich dynamics seen in Fig 5.6, where GS and ES are clearly reacting differently. This could be caused by Abuusa *et al.* investigating mainly 'free' relaxation oscillations, which appear during turn-on, while the modulation response is 'driven' by an external source. Furthermore, their frequency always increases with current. Yet, the GS-modulation enhancement seen in Fig 5.6 seems to vanish for high currents, where cut-off-frequencies saturate. This suggests that the relaxation oscillation damping factor is at least as important as the frequency itself, as damping would have to increase strongly to suppress resonances.

A closer look to the injection current region of Fig. 5.6 in question also reveals that the resonance frequency does not shift. This can be seen in Fig. 5.8, where the absolute modulation response is shown for GS (red) and ES (blue) versus frequency. For the panel on the left the device was simulated at a current slightly below ES threshold $J < J_{th}^{ES}$, so the GS is lasing, while the ES is not. Correspondingly, ES absolute modulation response is weak in comparison to GS response, as ES intensity is purely caused by spontaneous emission (cf. the light-current characteristic in

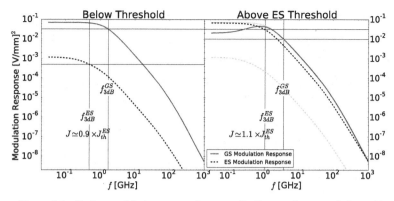

Figure 5.8: Absolute modulation response (not normalised) versus frequency f obtained by numerical simulation. ES (blue) and GS(red) modulation response is plotted on the same scale for currents $J < J_{th}^{ES}$ (left) and $J > J_{th}^{ES}$ (right.) Parameters for the numerical simulation as in Tab. 4, $\Delta J = 2 \times 10^{-7} \mathrm{enm}^{-2}\mathrm{ps}^{-1}$ and nonlinear scattering rates as in App. A.1.

Fig. 5.6). Both response curves exhibit no clear maximum, and ES cut-off frequency is relatively low at $f_{3dB}^{ES} < 1$ GHz.

Figure 5.8 (right) shows the absolute modulation responses, in exactly the same scaling as on the left, for a current slightly above ES lasing threshold $J \simeq 1.1 J_{th}^{ES}$. With the ES intensity being greatly enhanced by stimulated emission, absolute modulation is similarly strengthened. Contrary to this, the high frequency flank of the GS modulation response is almost left unchanged. For frequencies around $f = 1$ GHz a resonance peak can be seen. This feature is 'revealed' by the receding low-frequency side of the GS modulation response curve. There, the absolute GS modulation is slightly weaker as compared to the lower injection currents in Fig. 5.8 (a), despite the fact that GS lasing intensity has risen. This is most pronounced for the flat part of the modulation response curve towards low frequencies $f < 0.1$ GHz.

This is also the reason, why no normalisation was used for Fig. 5.8. The low-frequency modulation response is always used as the baseline when normalising. This would result in a 'lifting' of the response curve, and the resonance feature appears to 'grow' out of the modulation response curve, as opposed to the low-frequency side receding and the resonance feature being left standing. Due to this renormalisation the cut-off-frequency for the GS f_{3dB}^{GS} is so greatly enhanced in Fig 5.6: The cut-off frequency is defined as the modulation frequency for which response intensity has dropped to $\simeq 0.5$ compared to the *low-frequency modulation response*, so when only the low-frequency modulation response is reduced, the cut-off frequency shifts to higher values.

Interpreting the observations told in the previous paragraphs, however, is not straight-forward. It seems that ES lasing is suppressing the low-frequency modulations in the GS; possibly by preventing carrier differences to propagate through the relaxation scattering channel. However, fast modulations are able to reach GS levels regardless of the ES lasing state. Further investigations will therefore be presented.

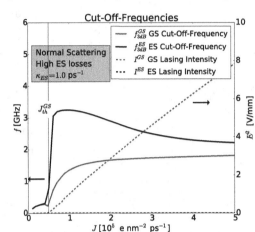

Figure 5.9: 3dB cut-off-frequencies for GS (red) and ES (blue) versus pump current obtained by numerical simulation. ES optical losses κ_{ES} were increased to prevent the ES from lasing. Consistently, Light-current characteristic of the simulation shown in dashed lines. Parameters for the numerical simulation as in Tab. 4, $\Delta J = 2 \times 10^{-7} \mathrm{enm}^{-2}\mathrm{ps}^{-1}$, $\kappa_{ES} = 1.0\mathrm{ps}^{-1}$ and nonlinear scattering rates as in App. A.1.

To verify, that it is really the onset of ES lasing that triggers this GS-modulation enhancement, Fig. 5.9 shows the GS and ES cut-off-frequencies and light-current characteristics with high optical losses in the ES. By setting $\kappa_{ES} = 1.0\mathrm{ps}^{-1}$, the ES was prevented from achieving a lasing state. As a result no sudden GS-modulation enhancement is visible after the onset of GS lasing, so it is clear that ES photons must play an important role in the GS-modulation enhancement.

On a side note, the sudden increase of ES modulation at GS threshold, already seen in the first figure of cut-off-frequencies Fig. 5.6, is still visible. But with no ES lasing threshold present, there is no minimum in the ES cut-off-frequency, as was seen for $J = J_{th}^{ES}$ in Fig. 5.6, just a decline towards higher frequencies. So there certainly seems to be also a modulation enhancement in reverse direction, albeit the ES is not lasing at that point.

For further investigation, Fig. 5.10 shows the 3dB cut-off-frequencies for GS (red) and ES (blue) versus pump current, similar to Fig. 5.6. Here, however, GS capture rates were turned off, $S_{b,in/out}^{GS,cap} = 0$. Therefore, with any direct interaction between GS and QW prevented, any modulation reaching the GS must have been transmitted through the cascade scattering chain QW-ES-GS, and therefore ultimately through the ES as an intermediate reservoir.

The similarities between Fig. 5.10 and Fig. 5.6, however, suggest that direct GS capture processes play no major role. The sudden enhancement of the GS modulation response at the onset of ES lasing is still present. Furthermore, the fact that almost nothing else changes can be attributed to the nature of the microscopically calculated scattering rates, i.e. the resulting GS capture rates are always slower than the cascade-scattering channel consisting of ES capture and relaxation.

Now, the opposite approach is shown in Fig. 5.11, where the relaxation scattering was turned off, $S^{rel} = 0$. This leads to a decoupling of the GS and ES and they

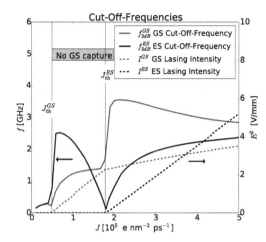

Figure 5.10: 3dB cut-off-frequencies for GS (red) and ES (blue) versus pump current obtained by numerical simulation. As opposed to Fig. 5.6, direct scattering from GS to QW were turned off. Light-current characteristic of the simulation shown in dashed lines. Parameters for the numerical simulation as in Tab. 4, $\Delta J = 2 \times 10^{-7} \mathrm{enm}^{-2}\mathrm{ps}^{-1}$ and linearised scattering rates as in App. A.2, with $S_{b,in/out}^{GS,cap} = 0$.

only indirectly interact through the reservoir carrier densities w_b. The strong GS-modulation enhancement feature is not present in this simulation. The overall shape of the cut-off-frequency curve is in agreement with simulations of QDs including only a single confined state [GIO11, LUE12].

Yet, there still remains some cross-influence between GS and ES. The ES cut-off frequency is clearly enhanced once the GS lasing threshold is crossed $J_{th}^{GS} < J < J_{th}^{ES}$, just like in the previous simulations of Fig. 5.6, 5.9 and 5.10. Furthermore, there seems to be a drastic enhancement of GS modulation, for a current in between GS and ES thresholds. This is, however, not the feature linked to the appearance of the ES. The individual modulation response curves (not shown here) look clearly different to Fig. 5.8.

To conclude this section, the anomalous GS cut-off-frequency increase was investigated. It has been shown to be linked to the onset of ES lasing and dependent on the relaxation scattering process, while it is not dependent on the direct capture from the QW. To the contrary, turning off the relaxation leads to the disappearance of this modulation enhancement.

From modulation response curves below and above ES thresholds, one can deduce that the ES acts like a high-pass filter, blocking only the low-frequency modulations from reaching the GS. This leads to the appearance of a resonance feature, enhancing the cut-off-frequency. It therefore seems that ES photons are acting as a buffer that reduce the modulation propagated to the GS.

5.6. Change of Cut-Off-Frequency with Carrier Loss Rates

Apart from the ES-lasing induced GS modulation enhancement studied in the previous section, there are other ways of increasing the maximum data transmission rate of QD-based devices. Two of these will shortly be covered in this section.

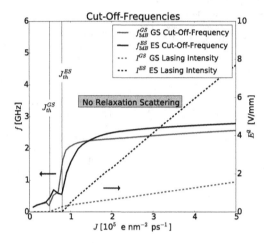

Figure 5.11: 3dB cut-off-frequencies for GS (red) and ES (blue) versus pump current obtained by numerical simulation. Light-current characteristic of the simulation shown in dashed lines. Relaxation from ES to GS was turned off, so that both carrier states are directly fed from the well. Parameters for the numerical simulation as in table 4, $\Delta J = 2 \times 10^{-7}\,\mathrm{enm^{-2}ps^{-1}}$ and linearised scattering rates as in App. A.2, with $S^{rel} = 0$.

Figure 5.12: 3dB cut-off-frequencies for GS (red) and ES (blue) versus pump current obtained by numerical simulation. Light-current characteristic of the simulation shown in dashed lines. Scattering rates were speed up by a factor of three, resulting in an increase of the cut-off-frequencies. Note the different scaling compared to previous figures, showing high-current dynamics where cut-off frequencies approach a constant value. Parameters for the numerical simulation as in Tab. 4, $\Delta J = 2 \times 10^{-7}\,\mathrm{enm^{-2}ps^{-1}}$ and nonlinear scattering rates as in App. A.1, multiplied with a factor of 3.

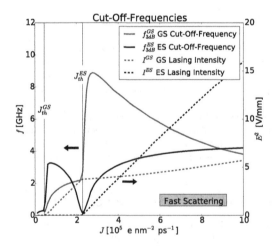

The most straight-forward way when simulating is by simply increasing the scattering rates. Fig. 5.12 shows the 3dB cut-off-frequencies for GS (red) and ES (blue) versus pump current obtained by numerical simulation for all scattering rates three times as fast. Note the different axis scaling compared to Fig. 5.6. The modulation response is faster, while all features are maintained. Also shown is the high-current range, for which both ES and GS cut-off-frequencies approach a constant value.

However, scattering rates cannot be simply increased in experiments, where they depend on QD shape, device structuring and material [BIM08], all of which have to already be controlled to yield correct telecommunication wavelengths and leave little room for further adjustments.

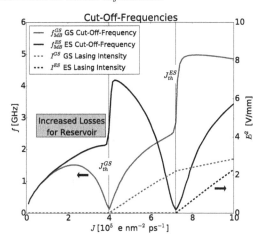

Figure 5.13: 3dB cut-off-frequencies for GS (red) and ES (blue) versus pump current obtained by numerical simulation. Light-current characteristic of the simulation shown in dashed lines. . Parameters for the numerical simulation as in Tab. 4, $\Delta J = 2 \times 10^{-7} \mathrm{enm}^{-2} \mathrm{ps}^{-1}$, $R^w_{loss} = 0.59 \mathrm{nm}^2 \mathrm{ps}^{-1}$ and nonlinear scattering rates as in App. A.1.

The interplay of carrier lifetime, mostly given by the combined loss term R^w_{loss} in the numerical model, with modulation response is complex [LIN12]. However, in the parameter range of this thesis, an increase of losses is predicted by Lingnau et al. to yield higher cut-off-frequencies. The results of a simulation with $R^w_{loss} = 0.59 \mathrm{nm}^2 \mathrm{ps}^{-1}$ is shown in Fig. 5.13. In accordance with the prediction, cut-off-frequencies are increased by $\sim 25\%$ in the GS and $\sim 40\%$ in the ES. However, this comes at the cost of higher threshold injection currents for both GS and ES.

5.7. Outlook for Modulation Response

To end this chapter, a last simulation shall be presented, highlighting the need for further work in this area. Figure 5.14 shows the cut-off-frequencies for GS and ES. Here, scattering rates were changed to lead to hole depletion and subsequent GS quenching by reducing hole capture rates to 5% as has been done in Sec. 4.2. The resulting cut-off-frequencies display rich dynamics. The individual modulation response curves for all currents were individually checked, to assure that no numerical error caused this behaviour.

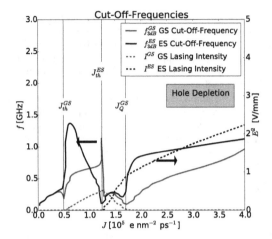

Figure 5.14: 3dB cut-off-frequencies for GS (red) and ES (blue) versus pump current obtained by numerical simulation. Scattering rates were changed to yield hole-depletion induced GS quenching. Light-current characteristic of the simulation shown in dashed lines. Parameters for the numerical simulation as in Tab. 4, $\Delta J = 2 \times 10^{-7} \mathrm{enm}^{-2}\mathrm{ps}^{-1}$ and linearised scattering rates as in App. A.2, hole captures reduced to 5%.

For one, the GS modulation cut-off frequency is *not* increased once two-state lasing starts, but apart from a small peak decreased. Additionally, overall performance has been greatly degraded with cut-off-frequencies well below 2GHz for most of the current range. However, the turned-off GS displays strong increase in the cut-off frequency for currents well above the GS quenching thresholds ($f \simeq 4.5$GHz), not seen in any other simulation so far. However, the GS is already turned off at that point.

Understanding all of these features, linking them to real and artificial causes of the parameters chosen, and examining the ones useful for data transmission is a topic suited for future research. In addition to pure numerical simulation of modulation response curves for even other scattering and parameter sets, there are several other points of interest.

The ES-lasing-induced GS modulation enhancement needs a verification of its suitability for data transmission. As seen form the individual modulation response curves, it is a suppression of low-frequency response that increases the 3dB cut-off-frequency. The high-frequency flank, however, stays almost constant when measured in absolute terms. This begs the question, whether a large signal analysis would actually yield decodable bits and needs to be studied.

Second, apart from the purely numerical study done here, a deeper, more physical understanding of the phenomenon needs to be developed. A good beginning could be deriving an analytical approximation. One might approximate the system as coupled, damped harmonic oscillators, of which only one is driven. Furthermore, the low-frequency range can probably be studied with steady-state simulations.

Lastly, the non-monotonous shape of cut-off-frequencies might be useful in designing experiments. As it appears that the GS modulation enhancement with ES lasing is linked to the relaxation scattering channel, it might be used to determine the strength of this scattering process in real QDs. Furthermore, if the cut-off frequency shape shown here for hole-depletion induced GS quenching can be generalised for a broad variety of parameters, measurements of the modulation responses could also be used to unambiguously determine the cause of GS quenching experimentally.

6. Pump-Probe Experiments

In this chapter two sets of experiments with a two-state QD device shall be described
and their respective behaviour will be reproduced by the numerical QD model. The
experiments were carried out by the work group of Prof. Woggon (TU Berlin), in
particular Yuecel Kaptan and Bastian Herzog.

6.1. Pump-Probe Setup

A principal investigative tool that can be used to measure carrier lifetimes in a
semiconductor device is the so called pump-probe setup. It will be explained in this
section. Basically, two different optical pulses of predefined shape, wavelength and
amplitude are injected into the sample that is to be studied. One of these is called
the 'pump pulse' and is meant to prepare the system in a specific state, e.g. revert
the system to optical transparency, while the second pulse is called the 'probe' and is
captured after it has travelled through the sample. Then analysing this probe pulse
yields information about the state of the system. Additionally, a tunable time delay
is introduced between the pump and the probe pulse, which allows a time-resolved
measurement of the systems response to the perturbation caused by the pump pulse.
A sketch is presented in Fig. 6.1.

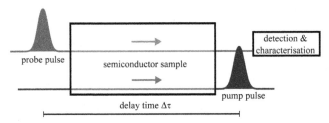

Figure 6.1: Sketch of a pump-probe experiment. Two pulses of predefined shape, wavelength
and energy are injected into the sample that is to be studied. The 'pump pulse' (blue) prepares
the state of the system, while the 'probe pulse' is analysed after it has travelled through the
prepared sample. By varying the time-delay $\Delta\tau$ between pump and probe pulse the time
evolution of the system can be extracted.

This technique has been used successfully by the AG Woggon in the past [DOM07,
GOM08, GOM10, MAJ11] and with the current level of sophistication they were able
to produce pulses with wavelengths between 900 and 2000 nm, spectral full-width-
at-half-maximum of 10-15 nm and achieve a time resolution of 250 fs.

6.2. Two-State Device Description

The device studied consists of 5 layers of self-assembled In(Ga)As QDs grown by
molecular beam epitaxy, which were overgrown each with another layer of InGaAs to
form a dot-in-a-well structure. The device was then processed with p and n-contacts,
and etched into a 1.33 mm long and 6 μm wide ridge wave-guide structure.

The front facet of the Fabry-Perot type resonator was simply cleaved and left unaltered, while the back facet was further processed in Israel by the group of Prof. Eisenstein at the Israel Institute of Technology. It was covered in a dichroic coating, which has a high reflectivity in the ES wavelength range and acts as an anti-reflective coating for the GS.

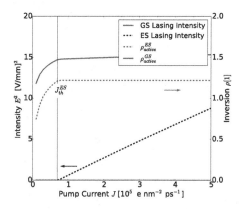

Figure 6.2: Simulated light-current characteristic of the device. Because of the dichroic coating, only ES lasing is achieved. Parameters as given in Tab. 4, with $\kappa_{GS} = 1.0\text{ps}^{-1}$, $R_{\text{loss}}^{W} = 0.1\text{nm}^2\text{ps}^{-1}$ and $J = 5 \cdot 10^{-5}\text{enm}^{-2}\text{ps}^{-1}$.

Therefore, the device behaves like a semiconductor optical amplifier on the GS, but is also able to achieve lasing on the ES. Fig. 6.2 shows the simulated light-current characteristic of the device (details of the simulation are explained in the following Sec. 6.3). Only ES lasing is achieved, while high GS losses prevent all GS lasing.

For this sample, the GS amplified spontaneous emission spectrum was centered around 1250 nm, while the ES lasing line was located at 1180 nm. The device was specifically designed for a variety of possible time-resolved and current dependent experiments to investigate the connection and scattering pathways of carriers between GS and ES.

6.3. First Experiment: Ground State Gain Recovery

In a preparatory step the different sub-ensembles of the inhomogeneously broadened GS spectrum were investigated. As the ES lasing line is sharp in contrast to the GS emission spectrum, it is clear that not all QDs participate in the ES lasing process. Yet, which parts of the broadened GS spectrum correspond to the active QDs is not easily accessible. The team of Prof. Woggon therefore initiated a series of dual-colour pump probe experiments: The pump pulse was centered at the peak of ES lasing intensity, while the probe pulse was stepwise varied from 1230 to 1300 nm.

With no injection current applied, the pump pulse would always be partially absorbed, increase the ES population and result in an increase in the GS population by scattering. The resulting change of absorption (= negative gain) was then measured with the probe pulse, centered on a specific part of the GS spectrum. The rise time of this GS gain increase was extracted for the different wavelength samples and a resonance was found. For the GS ensemble centered at 1270 nm response times were fastest and these QDs therefore correspond to the ES lasing subensemble. This corresponds to the optically active fraction of dots f^{act} in the numerical model of the QDs (see Sec. 2.3 for the description). They also deduced that the QDs centered at 1255 nm were sufficiently far away to be considered 'off-resonant', which translates to the inactive part of the QDs f^{inact} in the model.

In a second step this optically active GS QD ensemble at 1270 nm and the off-resonant QDs at 1255 nm were studied in more detail. A pump probe experiment was carried out for different injection currents with both pulses matching the respective wavelength. Time traces for this single-colour pump-probe experiment can be seen in Fig. 6.3. As can be expected, the gain recovery is faster for higher injection currents, because the scattering processes get faster and 2D reservoir occupations are higher.

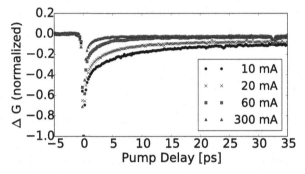

Figure 6.3: Time-resolved GS gain recovery (symbols) as measured by the work group of Prof. Woggon. These experimental GS gain curves were obtained via a single-colour pump-probe setup for a wavelength of 1255 nm. The time difference between pump and probe pulse is listed on the x-axis, while the gain response has been normalized. The fits were made by a sum of two exponential functions. The gain recovers faster for higher injection currents (different colours). See Fig. 6.4 for a simulated response. Redrawn after [KAP14b]

These curves were fitted with a two-exponential function with offset, corresponding to three different identified time domains. Y. Kaptan *et al.* attribute the sub-fs gain recovery time τ_1 to polarization dynamics. The second time scale τ_2 is on the order of several picoseconds and corresponds to the carrier-carrier scattering from ES and 2D-reservoir into the GS. Lastly, a long time scale was identified, with recovery times on the order of several nanoseconds. This was attributed to a refilling of an additional, only indirectly accessible carrier reservoir still above the well structure, e.g. sometimes called a separate confinement heterostructure (often abbreviated as SCH).

A simulation of such a pump-probe experiment is presented in Fig 6.4. To model the dichroic coating, the optical decay rate κ was split into two different decay rates. The GS light is subject to a high decay rate of $\kappa_{GS} = 1.0\text{ps}^{-1}$, accounting for the antireflective properties of the dichroic coating. Correspondingly the ES optical decay rate κ_{ES} was kept at the previous value of 0.05ps^{-1} to enable lasing. For a

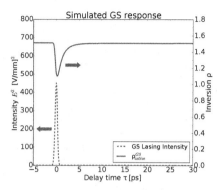

Figure 6.4: Simulated GS inversion ρ_{act}^{GS} (green line) versus time when a pump pulse has been injected on the GS (red dotted line). No probe pulse is needed for the simulation, as all state variables are easily extractable at all times. Only the recovery on the time scale of several picoseconds was modelled, as the polarization has been adiabatically eliminated and no separate confinement heterostructure is included. Parameters as given in Tab 4, except $\kappa_{GS} = 1.0\mathrm{ps}^{-1}$, $\kappa_{ES} = 0.05\mathrm{ps}^{-1}$, $R_{\mathrm{loss}}^{W} = 0.1\mathrm{nm}^2\mathrm{ps}^{-1}$ and $J = 5 \cdot 10^{-5}enm^{-2}\mathrm{ps}^{-1}$.

better agreement between experiment and simulation the bulk loss rate was changed to $R_{\mathrm{loss}}^{W} = 0.1\mathrm{nm}^2\mathrm{ps}^{-1}$. Furthermore, to account for the spectral width of the pump-pulse a stimulated emission term is added to the inactive QD GS-Eq. (2.65) with gain $g_{GS,ia} = 0.0575ps^{-1}$. No GS lasing is ever achieved with this choice of parameters, so there is no interaction of inactive and active QDs through GS light intensity and the distinction between these two subgroups is maintained on the basis of ES lasing.

Within the model used in this work only the picosecond time-scale was reproduced as both the ultrafast polarization dynamics and the long-term separate confinement heterostructure are not included. Note, that a perfect fit for this experiment was not attempted as the interpretation of the results led to a second set of experiments (see Sec. 6.4) and a closer collaboration between the AG Woggon and AG Schöll.

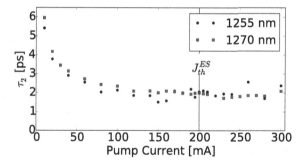

Figure 6.5: GS recovery time for single-colour pump-probe experiments carried out by Y. Kaptan *et al.* (TU Berlin). The time resolved response curves as seen in Fig. 6.3 were fitted with a two-exponential ansatz and the second time constant τ_2 was extracted. The subensembles participating in ES lasing (red squares) and the off-resonant QDs (blue circles) react identically to perturbations, even above the ES lasing threshold (vertical line). This time-scale corresponds to the intradot and/or QD-QW scattering processes. Redrawn after [KAP14b]

Fig. 6.5 shows the experimentally extracted time scales τ_2 versus injection currents. Opposite to their initial expectations, Y. Kaptan *et al.* found no difference between the off-resonant (1255 nm) and resonant (1270 nm) subensemble of the QDs. Additionally, they had hoped to identify the dominating scattering path for the GS recovery. The two possible pathways are either a direct QW-QD capture ($S_{b,in}^{GS,cap}$ in the numerical model) or a cascade from QW to ES and then GS (corresponding to $S_{b,in}^{rel}$ and $S_{b,in}^{GS,cap}$ in the model). Yet, a numerical simulation with the model presented in this work proved, sadly, that distinguishing these processes was not possible with the type of experiment employed by Y. Kaptan *et al.*.

A simulation of GS recovery for different currents and scattering schemes was performed and the results are shown in Fig. 6.6. For the first scattering scheme, the direct capture was turned off via setting $S_{b,in}^{GS,cap} = 0$, so that all carriers reach the GS through the cascade scattering process QW-ES-GS. The rest of the scattering rates were left as presented in Sec. 2.3.2 and shown in App. A.2 (red line in Fig. 6.6). Secondly, the direct capture was turned back on and the full set of microscopically calculated scattering equations was used, corresponding to a mixed capture (green line in Fig. 6.6). Lastly, the direct capture channels $S_{b,in}^{GS,cap}$ were sped up by a factor of 5, and the cascade participating scattering rates $S_{b,in}^{rel}$ and $S_{b,in}^{GS,cap}$ slowed down to 20%, resulting in a direct-capture dominated scattering scheme(black line in Fig. 6.6).

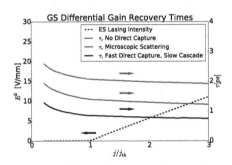

Figure 6.6: Simulated GS inversion recovery time scales τ (solid lines) versus normalized injection current j and light intensity of the ES (dashed blue line). Three different sets of scattering rates were used. Green: linearised scattering rates as calculated in Sec. 2.3.2 and shown in App. A.2. Red: Direct capture $S_{b,in}^{GS,cap}$ was turned off. Black: Direct capture $S_{b,in}^{GS,cap}$ speed up by a factor 5 and the cascade process ($S_{b,in}^{rel}$ and $S_{b,in}^{GS,cap}$) was slowed down to 20%. Note: magnitude can be tuned as a whole to reproduce experimentally measured time constants (seen in Fig.6.5). Parameters as given in Tab. 4, with $\kappa_{GS} = 1.0\text{ps}^{-1}$, $R_{loss}^W = 0.1\text{nm}^2\text{ps}^{-1}$ and $J = 5 \cdot 10^{-5}\text{enm}^{-2}\text{ps}^{-1}$.

Figure 6.6 shows the resulting time scales versus normalized pump currents. The individual magnitude of the three curves is of small importance, as each of them could be tuned as a whole without changing the scattering pathways by multiplying all scattering channels by a constant factor. The overall shape, however, nicely reproduces the experimental findings seen in Fig. 6.5. The underlying process can, unsurprisingly, be summed up as 'higher injection current increases the gain recovery'. Yet, there is no qualitative difference between the different scattering schemes (red, green and black curves) and therefore distinguishing which pathway is active is not possible from the experimental data obtained.

Furthermore, the onset of ES lasing (see blue dotted line in Fig. 6.6) is never clearly visible in the recovery time scales τ, not even for the pure cascade scattering scheme (red line). The dominating scattering pathway could therefore not be identified. The numerical model, however, proved capable of reproducing the experimental findings.

Additionally, while studying the response dynamics of the system via simulation, an intriguing feature was observed: Through cascade scattering processes the perturbation of the GS population gets translated into a perturbation of ES carriers. So when the GS loses population by amplifying the incoming pump pulse, ES carriers are also decreasing. Naturally, this leads to a drop in the ES lasing intensity and induces relaxation oscillations in the ES intensity while recovering. These oscillations become especially pronounced close to the ES lasing threshold.

Having already envisioned a similar idea, the group of Prof. Woggon set out to try and reproduce this numerically predicted behaviour. This was done in a second experiment and will be presented in the following section.

6.4. Second Experiment: Excited State Intensity Recovery

After the injection of the pump pulse on the GS, the perturbation of GS carriers is ultimately transformed into a drop of ES intensity. Figure 6.7 shows the ES intensity versus time as simulated by the numerical model, after a GS pulse reduced GS inversion back to transparency at $\tau = 0$. The recovery of this ES intensity appears on the order of several hundred picoseconds and is greatly delayed when compared to the carrier recovery seen in Fig. 6.4, which takes less than 20 ps. Furthermore, the ES intensity first drops by about 30% and then overshoots while recovering. The strength and time scale of this oscillatory behaviour is dependent on the injection current and strongest in the vicinity of the ES lasing threshold.

The experimental setup was changed to study the ES intensity. Instead of using a probe pulse, the outgoing ES intensity was constantly monitored by a streak camera with temporal resolution of 30 ps. With this time-resolved measurement the ES-intensity recovery-curves were obtained.

Figure 6.8 (a) shows the ES intensity drop and recovery curves for different injection currents J. The solid lines are the experimentally measured data, whereas the dotted lines are the best fit obtained with the numerical model by changing R_{loss}^w, g_{ES}, κ_{ES} and N^{QD}, all of which are parameters which are different from device to device and therefore have to be inferred or measured. To account for the temporal resolution, the simulation was convoluted with a Gaussian profile with a full-width-at-half-maximum of 30 ps. The threshold current of the device was about 190 mA and the injection current of the simulation was normalised with respect to this value. The black sample represents the only sub-threshold measurement.

Just as predicted by the numerical simulations, the ES recovers on the order of several hundred picoseconds. Thanks to some assistance provided by Benjamin Lingnau, the fit agrees almost perfectly with the experimental data. The parameters are given in Tab. 7.

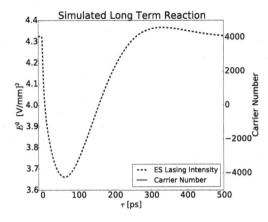

Figure 6.7: Reaction of ES intensity versus time, after a pulse is injected on the GS at $\tau = 0$. The ES intensity first drops by about 30% and then overshoots while recovering. Note the larger time scales compared to Fig. 6.4. Parameters as given in Tab. 4, except $\kappa_{GS} = 1.0\text{ps}^{-1}$, $\kappa_{ES} = 0.05\text{ps}^{-1}$, $R_{loss}^{W} = 0.1\text{nm}^2\text{ps}^{-1}$ and $J = 5 \cdot 10^{-5}\text{enm}^{-2}\text{ps}^{-1}$.

The recovery takes longest, when the current is close to the ES threshold J_{th}^{ES} (190mA, brown curve in Fig. 6.8), where the relative drop in intensity is also the most pronounced. To illustrate the behaviour, the current-dependent dynamics are shown in Fig. 6.8 (b), where the ES minimum time (blue) and depth (red) are plotted versus injection current J. The squares represent the values obtained from the six experimentally studied recovery curves in (a), while the lines were obtained by numerical simulation. The ES intensity recovery is slowest at the ES lasing threshold, and speeds up towards the flanks. Below a critical pump current, the simulation predicts an almost instantaneous ES drop off. However, for that current range no experimental recovery was measured, due to the low ES intensity and limited time-resolution of the streak-camera.

The overall shape of the ES recovery time scale and minimum depth is linked to the dampening and frequency of relaxation oscillations. When the system is perturbed by the pump pulse, it is driven out of the fixed point and relaxes back to equilibrium once all of the GS pulse intensity has passed. This is similar to a turn-on scenario, where the system also transiently approaches the steady state and therefore also induces relaxation oscillations. Hence, the depth and timing of the minimum are linked to the dampening and frequency of relaxation oscillations, which in turn correspond to the real and imaginary part if the partaking eigenvalue.

As has been shown, the LI-curve with the onset of lasing can be modelled, in a first approximation, by a transcritical bifurcation where the 'off' and 'on' solutions exchange stability at the threshold current (see Sec. 2.1.3). The damping of the relaxation oscillations is therefore weakest close to the threshold and increases for values far away from it. The imaginary part, however, is not as easily obtained and must be derived from a model with at least two state variables where the lasing state is actually a stable focus.

To further underline the nature of these oscillations, Fig. 6.9 plots the ES intensity

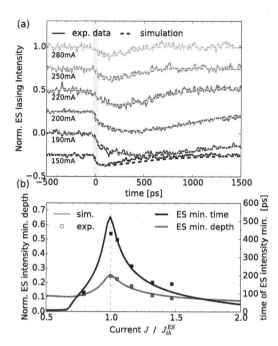

Figure 6.8: (a) ES intensity reaction after a pump pulse on the GS. Measured data: Solid lines. Simulated data: Dashed lines. The curves are plotted with an offset for different pump currents. Only the relative drop in the ES intensity was measured and simulation results were therefore normalized as well. The time and relative depth of the minimum were evaluated and are plotted in (b). Squares represent the six experimentally measured curves shown in (a) and the solid line the numerically simulated values. The injection current was normalized to the ES threshold current J_{th}^{ES}. The parameters are given in Tab. 7. Redrawn after [KAP14b].

fluctuation together with the change of average electron number per QD Δn_e:

$$n_e = \frac{w_e}{N^{QD}} + f^{act}(2\rho_e^{GS} + 4\rho_e^{ES}) + f^{inact}(2\rho_{e,ia}^{GS} + 4\rho_{e,ia}^{ES}). \qquad (6.1)$$

The current is $J = 200$ mA. After an ultra-fast recovery shortly after the pump pulse, the carrier number (green line) exhibits oscillatory behaviour on the same time-scales as the ES intensity (blue line). This is a clear indicator for relaxation oscillations and results from the interacting charge carriers and lasing fields periodically exchanging energy. Even for this current close to the threshold oscillations are quite damped, as is typical for QD lasers [BIM99], so that no transient overshoot is ever produced.

Additionally, Fig. 6.10 displays the results for an excitation of inactive QD subensembles by shifting the GS pulse wavelength either to lower (red, 1230 nm) or higher (blue, 1290) wavelengths. The resulting minimum is in both cases shallower than for resonant excitation (green, 1260 nm) while the overall shape is maintained. Most importantly, the timing of the minimum is nearly unchanged, indicating that the underlying mechanics are the same as for resonant excitation and that relaxation oscillations are induced in a similar fashion. This is also what is expected, as the system redistributes charge carriers, even across the QW barrier, from inactive to active subensembles on the time scales of several picoseconds. The dent in the

Table 7: Parameters used in the calculations of the experimental fits. Parameters not given here are the same as in Tab. 4.

Symbol	Value	Meaning
T	300K	Temperature
g_{GS}	0.05ps^{-1}	GS linear gain, active QD
g_{GS}	0.025ps^{-1}	GS linear gain, inactive QD
g_{ES}	0.1ps^{-1}	ES linear gain, active QD
κ_{GS}	1.0ps^{-1}	Optical GS losses, anti-reflective
κ_{ES}	0.068ps^{-1}	Optical ES losses, reflective
β	1×10^{-2}	Spontaneous emission factor
Z^{QD}	1.5×10^7	Number of QDs
N^{QD}	$5 \times 10^{10}\text{cm}^{-2}$	Area density of QDs
f^{act}	0.5	Fraction of active dots
R^W_{loss}	$0.04\text{nm}^2\text{ps}^{-1}$	QW loss rate

overall number of carriers is simply reduced by the fact that the inactive QDs are spectrally more diverse and therefore absorb less of the GS pulse intensity. Hence, the resulting recovery curve is identical in shape but smaller in amplitude to the resonant excitation.

The simulation of this off-resonant excitation was achieved by setting the ES gain of the active QDs to zero $g_{GS} = 0$. The resulting time-dependent recovery curves (Fig. 6.10, dotted lines) once again nicely reproduce the experimentally obtained data (solid lines). Note, that apart from the gain and current J the same parameters were used for all simulations of this experiment, highlighting that the numerical model reproduces the devices dynamic over a broad range of experimental set-ups.

Overall, both experiments and the cooperation with the group of Prof. Woggon have proven to be a great success. After a *prediction* by the numerical model the behaviour was observed in the experiment. The subsequent modelling of the ES intensity recovery-curves lead to a satisfactory result. Not only is each individual curve in excellent agreement with the experiment, but the current dependence is also nicely reproduced. It is therefore evident that the numerical model used is capable of describing even such time-dependent dynamics for the interaction of GS and ES carriers and light fields.

The experimental details and additional information for this section are published in APL as part of Ref. [KAP14b].

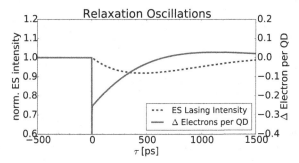

Figure 6.9: Simulated ES intensity (blue) and electron number per QD (green) reaction to a pump pulse on the GS at $\tau = 0$ for $J = 200$ mA. As typical for relaxation oscillations the carrier recovery exhibits the same time-scale as the simultaneous intensity fluctuation on the ES. The experimental trace of the ES minimum depth and timing of Fig. 6.8 (b) can therefore be identified as resulting from the current dependence of the relaxation oscillation frequency and damping. The parameters used are given in Tab. 7.

Figure 6.10: ES intensity reaction after a pump pulse on the GS for excitation of resonant ('active') QD subensembles (green) and non-resonant ('inactive') QD subensembles (red, blue) at $J = 200$ mA. The measured data (solid lines) were also simulated with the numerical model (dashed lines). Non-resonant excitation was modelled via setting $g_{GS} = 0$, resulting in no absorption of the pump pulse by active QDs. The rest of the parameters are given in Tab. 7.

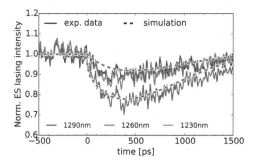

7. Summary and Outlook

In the scope of this work the dynamic properties of quantum dot (QD) lasers were numerically simulated, with the focus on two-state devices. These types of semiconductor lasers can simultaneously achieve lasing on two separate wavelengths, owing to the discrete set of energy levels inside the QD. The lasing states are called ground state (GS) lasing and excited state (ES) lasing, referring to the confined QD level that is involved.

From semiconductor and general laser properties, the derivation of the semiclassical laser-equations was presented. Based on previous works these laser equations were modified to include GS and ES electric fields and carrier-carrier scattering rates were taken as calculated by B. Lingnau and N. Majer.

After reproducing the known results of single-state lasing with this numerical model, the first step in investigating two-state lasing was taken. The current-dependent steady states of two-state lasing QD lasers were simulated and presented in light-current characteristics. These were then compared to the experimental findings of Markus *et al.* [MAR03, MAR03c] and the importance of incomplete gain clamping could be seen.

As a next step, the GS quenching was investigated. After presenting the possible explanations identified in the literature, namely homogeneous broadening increase, self-heating and electron-hole dynamics, the numerical model was used to reproduce these claims. However, after taking into account the latest experimental findings and weighing in some of the arguments brought forward, the asymmetric electron-hole dynamics emerged as the dominating cause. Based on this, an analytical approximation for the GS and ES lasing regimes was derived and applied, to visualize the carrier dynamics during GS quenching.

As a result of the analytical derivations, the key parameters influencing GS quenching were identified. Consequently, these parameters were used in 2D-plots of the lasing regimes. GS quenching is strongly linked to the asymmetric electron-hole dynamics and can only be reproduced by achieving hole depletion. Furthermore, it always constitutes a transition region in parameter space, inbetween purely ES lasing devices and stable two-state lasing regions. This could potentially be exploited when experimentally trying to find the GS quenching regions, and the few experimental data available so far is in good agreement with the 2D-plots. On a side note, the turn-on dynamics of GS quenching devices were also presented. GS lasing and ES lasing can be temporarily observed even for currents outside their stable lasing regimes.

Then, the modulation response of two-state lasing was simulated. As a result of the interaction of both active lasing modes, a drastic increase in the GS cut-off-frequency was observed. The origin of this striking feature could be traced to the onset of ES lasing and the coupling of both GS and ES carriers through the relaxation scattering. However, no experimental data on this type of device was available and more work needs to be done in this direction.

Lastly, a pump-probe experiment of a two-state lasing device by the group of Prof. Woggon was numerically reproduced. Furthermore, the predictions of the numerical model could be verified in a section set of experiments.

Overall, the numerical model has proven itself to be consistent with available data. It is capable of describing the main features, including two-state lasing and ground-state quenching. However, with the scattering scheme as currently calculated, no hole-depletion was achieved. This could be rectified by recalculating the scattering rates from a different setup of energy separations. Furthermore, one could also think of including additional hole states to improve the set of differential equations even further.

Part of this work was submitted as [ROE14] to IEEE Journal of Quantum Electronics.

Appendices

A. Scattering Rates

A.1. Fully Non-Linear Rates

With charge carrier densities in the surrounding quantum well w_e and w_h given in $[10^{15} \text{ m}^{-2}]$, the capture rates $S_{b,in}^{m,cap}$ in $[\text{ps}^{-1}]$ with $b \in \{e, h\}$ and $m \in \{GS, ES\}$ are given by:

$$S_{b,in}^{m,cap} = \frac{(A_1 w_e^2 + A_2 w_h^2)\exp(C_1 w_e + C_2 w_h)}{1 - B_1 w_h/w_e + B_2 w_e + B_3 w_h - B_4 w_e^2 + B_5 w_h^2 + B_6 w_e w_h}. \quad (A.1)$$

Table 8: GS capture scattering parameters

$S_{e,in}^{GS,cap}$		$S_{h,in}^{GS,cap}$	
Par.	Value	Par.	Value
A_1	0.00895343	A_1	0.0000743182
A_2	0.0000926157	A_2	0.00177834
B_1	-0.0439039	B_1	-0.00545883
B_2	0.211373	B_2	0.0108673
B_3	0.194881	B_3	0.153377
B_4	-0.00985679	B_4	0.000844681
B_5	0.00207208	B_5	-0.000207441
B_6	0.0170416	B_6	0.00273418
C_1	-0.0114633	C_1	0.0162571
C_2	0.0116515	C_2	0.000169687

Table 9: ES capture scattering parameters

$S_{e,in}^{ES,cap}$		$S_{h,in}^{ES,cap}$	
Par.	Value	Par.	Value
A_1	0.055805	A_1	0.00188448
A_2	-0.000832346	A_2	0.0127417
B_1	-0.156698	B_1	0.256447
B_2	0.908605	B_2	0.170684
B_3	-0.069774	B_3	0.448176
B_4	-0.0184423	B_4	0.00634063
B_5	-0.00705115	B_5	0.0121547
B_6	0.0595145	B_6	-0.00479526
C_1	-0.0173271	C_1	-0.000115901
C_2	0.00999964	C_2	0.00950479

Relaxation scattering rates are given by:

$$S_{b,in}^{Rel} = \frac{(A_1 w_e + A_2 w_h)\exp(C_1 w_e + C_2 w_h)}{1 - B_1 w_h / w_e + B_2 w_e + B_3 w_h - B_4 w_e^2 + B_5 w_h^2 + B_6 w_e w_h}. \tag{A.2}$$

Table 10: Relaxation scattering parameters

$S_{e,in}^{rel}$		$S_{h,in}^{rel}$	
Par.	Value	Par.	Value
A_1	0.493054	A_1	0.281967
A_2	0.138407	A_2	0.689836
B_1	0.0401102	B_1	-0.0124172
B_2	0.0641796	B_2	0.22247
B_3	0.384811	B_3	0.197314
B_4	0.00835259	B_4	-0.00197186
B_5	0.00746199	B_5	0.0035868
B_6	0.0129023	B_6	0.00255207
C_1	-0.018721	C_1	0.00295926
C_2	0.0165946	C_2	0.00306085

A.2. Linearised Size-Dependent Scattering Rates

The size-dependent scattering rates were derived from the microscopically calculated ones of the previous section. r is a scaling parameter ranging from 0.5 to 2, with $r = 1$ aligned to the QDs with energy spacing as described in Sec. 2.3:

$$
\begin{aligned}
S_{e,in}^{GS,cap} &= 0.016 \cdot r^{-1.48} w_e \\
S_{h,in}^{GS,cap} &= 0.0108 \cdot r^{-1.40} w_h \\
S_{e,in}^{ES,cap} &= 0.032 \cdot r^{-1.50} w_e \\
S_{h,in}^{ES,cap} &= 0.0186 \cdot r^{-1.33} w_h \\
S_{e,in}^{Rel} &= \frac{0.88 \cdot r^{-0.47} w_e}{0.93 \cdot r^{1.3} + w_e} \\
S_{h,in}^{Rel} &= \frac{2.2 \cdot r^{-0.50} w_h /}{2.27 \cdot r^{0.87} + w_h}
\end{aligned}
\tag{A.3}
$$

B. Deutsche Zusammenfassung und Ausblick

In dieser Arbeit wurden Quantenpunktlaser numerisch und analytisch untersucht. Der Fokus lag dabei auf Quantenpunktlaser, die simultan auf zwei unterschiedlichen Wellenlängen emittieren. Dies ist nur durch die besondere energetische Struktur der Quantenpunkte möglich, deren diskretes Energiespektrum die gleichzeitige Inversion von Grund- und erstem angeregten Zustand erlaubt.

Zunächst wurden grundlegende Laserbestandteile und Eigenschaften zusammen mit eine kurzen Überblick ihrer Historie vorgestellt, von wo aus im weiteren Verlauf die semiklassischen Lasergleichungen hergeleitet wurden. In diesem Differentialgleichungssystem werden die quantemechanischen Zustände des aktiven Lasermediums mit den klassischen Maxwellgleichungen gekoppelt und mit ihm lassen sich eine Fülle von Halbleiterlaserszenarien untersuchen. Die Ladungsträgermechanik wird dabei neben der stimulierten Emission, auch von den Streuraten zwischen verschiedenen Zuständen dominiert. Diese Streuraten wurden in früheren Arbeiten bereits berechnet und die Ergebnisse von N. Majer und B. Lingnau wurden für diese Arbeit aufgegriffen und verwandt.

Nachdem mit dem semiklassischen Lasermodell zur Modellvalidierung das Verhalten von Laser mit lediglich einem aktiven optischen Übergang reproduziert wurde, wurden zunächst die Fixpunkte und die Laserkennlinie für Zwei-Wellenlängen-Quantenpunktlaser berechnet. Dabei zeigte sich, dass die berechneten Kennlinien in gutem Einklang mit den bereits veröffentlichten Experimenten von Markus *et al.* sind. Insbesonderen konnte nun auch die bereits heuristisch motivierte unvollständige Ladungsträgersättigung numerisch reproduziert werden, die es dem energetisch höheren angeregten Zustand erlaubt Ladungsträger anzusammeln.

Im Folgenden wurde dann eine besondere Eigenschaft der Zwei-Wellenlängen-Laser näher untersucht: der Grundzustandsintensitätsabfall für steigende Pumpströme. Dabei wurden die verschiedenen Mechanismen aus aktuellen wissenschaftlichten Publikationen vorgestellt, und mit Hilfe des numerischen Quantenpunktmodells reproduziert. Neben Selbsterhitzungsprozessen oder einer Vergrößerung der homogenen Linienbreite, stellte sich jedoch im Laufe dieser Aufarbeitung heraus, dass das asymmetrischen Elektron-Loch-Verhalten der treibende Faktor hinter diesem anormalen Intensitätsabfall ist. Darauf aufbauend wurden mit Hilfe einer analytischen Näherung die Laserbedingungen für den Grundzustand und angeregten Zustand hergeleitet, die nachfolgend zur Visualisierung der Ladungsträgerprozesse benutzt wurden.

Mit den Ergebnissen dieser analytischen Studien konnten dann die wichtigen Parameter identifiziert und deren Einfluss in 2D-Grafiken aufgetragen werden. Dabei wurden insbesondere die Variation der Laserintensitäten mit den optischen Verlusten, der Temperatur und der Quantenpunktgröße untersucht. Der Grundzustandsintensitätsabfall tritt dabei nur in einer relativ eng begrenzten Zone des hochdimensionalen Parameterraums auf, und stellt ein Übergangsphänomen zwischen Laser, die nur auf dem Grundzustand abstrahlen, und Lasern, die nur auf angeregten Zustand abstrahlen, dar.

Im Folgenden wurden dann die Modulationseigenschaften dieser Quantenpunktlaser simuliert, wie sie z.B. für die Datenübertragung von Bedeutung sind. Dabei wurde ein deutlicher Geschwindigkeitszuwachs für den Grundzustand vorhergesagt, sobald der angeregte Zustand ebenfalls aktiv wird. Der Ursprung dieses Übergängs konnte dabei mit dem Relaxationsprozess der Ladungsträger in Verbindung gebracht werden. Da aber jedoch bisher keinerlei experimentelle Daten zu diesem Thema veröffentlich wurden, sind weitere Untersuchungen erforderlich.

Im letzten Abschnitt wurde dann in Kooperation mit der Arbeitsgruppe von Prof. Woggon (TU Berlin) die Ladungsträgerdynamik experimentell untersucht. Dabei konnte das Quantenpunktmodell nicht nur bereits aufgenommene Daten reproduzieren, sondern auch aktiv Vorhersagen treffen, die dann in einem zweiten Experiment bestätigt wurden.

Insgesamt hat sich das Modell als äußerst verlässlich erwiesen, sowohl was die Kennlinien als auch die Zeitentwicklung des Systems angeht. Weitere Verbesserungen könnte man erreichen, indem man zusätzliche angeregte Zustände, insbesondere für die Löcher, integriert. Dabei könnten auch die Streuraten neu berechnet werden und sowohl die Phonon-Streuung als auch nicht-parabolische Wellenfunktionen berücksichtigt werden.

References

[ABU13] M. Abusaa, J. Danckaert, E. A. Viktorov, and T. Erneux: *Intradot time scales strongly affect the relaxation dynamics in quantum dot lasers*, Phys. Rev. A **87**, 063827 (2013).

[ALE07] R. R. Alexander, D. Childs, H. Agarwal, K. M. Groom, H. Y. Liu, M. Hopkinson, and R. A. Hogg: *Zero and controllable linewidth enhancement factor in p-doped 1.3 μm quantum dot lasers*, Jpn. J. Appl. Phys. **46**, 2421 (2007).

[ARA82] Y. Arakawa and H. Sakaki: *Multidimensional quantum well laser and temperature dependence of its threshold current*, Appl. Phys. Lett. **40**, 939 (1982).

[ARS13] D. Arsenijević, M. Kleinert, and D. Bimberg: *Phase noise and jitter reduction by optical feedback on passively mode-locked quantum-dot lasers*, Appl. Phys. Lett. **103**, 231101 (2013).

[ARS14] D. Arsenijević, A. Schliwa, H. Schmeckebier, M. Stubenrauch, M. Spiegelberg, D. Bimberg, V. Mikhelashvili, and G. Eisenstein: *Comparison of dynamic properties of ground- and excited-state emission in p-doped InAs/GaAs quantum-dot lasers*, Appl. Phys. Lett. **104**, 181101 (2014).

[ASA86] M. Asada, Y. Miyamoto, and Y. Suematsu: *Gain and the threshold of three-dimensional quantum-box lasers*, IEEE J. Quantum Electron. **22**, 1915–1921 (1986).

[BAS61] N. G. Basov: *Possibility of using indirect transitions to obtain negative temperatures in semiconductors*, J. Exp. Theo. Phys. **39**, 1033 (1961).

[BIM99] D. Bimberg, M. Grundmann, and N. N. Ledentsov: *Quantum Dot Heterostructures* (John Wiley & Sons Ltd., New York, 1999).

[BIM08] D. Bimberg: *Quantum dot based nanophotonics and nanoelectronics*, Electron. Lett. **44**, 168 (2008).

[BIM08a] D. Bimberg: *Semiconductor Nanostructures* (Springer, Berlin, 2008).

[BIM12] D. Bimberg: *Vom hässlichen Entlein zum Schwan- vor fünfzig Jahren wurde der Halbleiterlaser erfunden.*, Physik Journal **Mai** (2012).

[BON11] R. Bonk, T. Vallaitis, J. Guetlein, C. Meuer, H. Schmeckebier, D. Bimberg, C. Koos, and J. Leuthold: *The input power dynamic range of a semiconductor optical amplifier and its relevance for access network applications*, IEEE Photon. J. **3**, 1039–1053 (2011).

[CAO09] Q. Cao, S. F. Yoon, C. Z. Tong, C. Y. Ngo, C. Y. Liu, R. Wang, and H. X. Zhao: *Two-state competition in 1.3-μm multilayer InAs/InGaAs quantum dot lasers*, Appl. Phys. Lett. **95**, 191101 (2009).

[CHO99] W. W. Chow and S. W. Koch: *Semiconductor-Laser Fundamentals* (Springer, Berlin, 1999).

[CHO13] C. U. Choe, H. Jang, H. M. Ri, T. Dahms, V. Flunkert, P. Hövel, and E. Schöll: *Simultaneous stabilization of periodic orbits and fixed points in delay-coupled Lorenz systems*, Cybernetics and Physics **1**, 155–164 (2012).

[DIN76] R. Dingle and C. H. Henry: *Quantum effects in heterostructure lasers*. United States Patent No. 3982207 (1976).

[DOK12] N. Dokhane, G. P. Puccioni, and G. L. Lippi: *Slow dynamics in semiconductor multi-longitudinal-mode laser transients governed by a master mode*, Phys. Rev. A **85**, 043823 (2012).

[DOM07] S. Dommers, V. V. Temnov, U. Woggon, J. Gomis, J. Martinez-Pastor, M. Lämmlin, and D. Bimberg: *Complete ground state gain recovery after ultrashort double pulses in quantum dot based semiconductor optical amplifier*, Appl. Phys. Lett. **90**, 033508 (2007).

[DRZ10] L. Drzewietzki, G. A. P. The, M. Gioannini, S. Breuer, I. Montrosset, W. Elsäßer, M. Hopkinson, and M. Krakowski: *Theoretical and experimental investigations of the temperature dependent continuous wave lasing characteristics and the switch-on dynamics of an InAs/InGaAs quantum-dot semiconductor laser*, Opt. Commun. **283**, 5092–5098 (2010).

[EIN17] A. Einstein: *Zur Quantentheorie der Strahlung*, Phys. Z. **18**, 121–128 (1917).

[ELI14] P. G. Eliseev: *Fiftieth anniversary of diode lasers: early history at Lebedev Institute*, Semicond. Sci. Technol. **27**, 1–6 (2014).

[ERN10b] T. Erneux and P. Glorieux: *Laser Dynamics* (Cambridge University Press, UK, 2010).

[FLU07] V. Flunkert and E. Schöll: *Suppressing noise-induced intensity pulsations in semiconductor lasers by means of time-delayed feedback*, Phys. Rev. E **76**, 066202 (2007).

[GOM08] J. Gomis-Bresco, S. Dommers, V. V. Temnov, U. Woggon, M. Lämmlin, D. Bimberg, E. Malić, M. Richter, E. Schöll, and A. Knorr: *Impact of Coulomb scattering on the ultrafast gain recovery in InGaAs quantum dots*, Phys. Rev. Lett. **101**, 256803 (2008).

[GOM10] J. Gomis-Bresco, S. Dommers-Völkel, O. Schops, Y. Kaptan, O. Dyatlova, D. Bimberg, and U. Woggon: *Time-resolved amplified spontaneous emission in quantum dots*, Appl. Phys. Lett. **97**, 251106 (2010).

[GIO06] F. S. Giorgi, G. Lazzeri, G. Natale, A. Iudice, S. Ruggieri, A. Paparelli, L. Murri, and F. Fornai: *MDMA and seizures: a dangerous liaison?*, Ann. N. Y. Acad. Sci. **1074**, 357–364 (2006).

[GIO11] M. Gioannini and M. Rossetti: *Time-domain traveling wave model of quantum dot DFB lasers*, IEEE J. Sel. Top. Quantum Electron. **17**, 1318–1326 (2011).

[GIO12] M. Gioannini: *Ground-state quenching in two-state lasing quantum dot lasers*, J. Appl. Phys. **111**, 043108 (2012).

[GRE13] D. Gready, G. Eisenstein, M. Gioannini, I. Montrosset, D. Arsenijević, H. Schmeckebier, M. Stubenrauch, and D. Bimberg: *On the relationship between small and large signal modulation capabilities in highly nonlinear quantum dot lasers*, Appl. Phys. Lett. **102**, 101107 (2013).

[GRU97] M. Grundmann and D. Bimberg: *Theory of random population for quantum dots*, Phys. Rev. B **55**, 9740 (1997).

[HAK85] H. Haken: *Light, Vol. 2* (North-Holland, Amsterdam, 1985).

[HAK86] H. Haken: *Laser Light Dynamics*, vol. II (North Holland, 1st edition edition, 1986).

[HOF11] W. Hofmann, M. Müller, P. Wolf, A. Mutig, T. Gründl, G. Böhm, D. Bimberg, and M. Amann: *40 Gbit/s modulation of 1550 nm VCSEL*, Electron. Lett. **47**, 270–271 (2011).

[JI10] H. M. Ji, T. Yang, Y. L. Cao, P. F. Xu, and Y. X. Gu: *Self-heating effect on the two-state lasing behaviors in 1.3-µm InAs-GaAs quantum-dot lasers*, Jpn. J. Appl. Phys. **49**, 072103 (2010).

[JIN08] C. Y. Jin, H. Y. Liu, Q. Jiang, M. Hopkinson, and O. Wada: *Simple theoretical model for the temperature stability of InAs/GaAs self-assembled quantum dot lasers with different p-type modulation doping levels*, Appl. Phys. Lett. **93**, 161103 (2008).

[KAP14b] Y. Kaptan, A. Röhm, B. Herzog, B. Lingnau, H. Schmeckebier, D. Arsenijević, V. Mikhelashvili, O. Schops, M. Kolarczik, G. Eisenstein, D. Bimberg, U. Woggon, N. Owschimikow, and K. Lüdge: *Stability of a quantum dot excited state laser during simultaneous ground state amplification*, Appl. Phys. Lett. **105**, 191105-1–191105-4 (2014).

[KIM10f] Y. J. Kim, Y. K. Joshi, and A. G. Fedorov: *Thermally dependent characteristics and spectral hole burning of the double-lasing, edge-emitting quantum-dot laser*, J. Appl. Phys. **107**, 073104 (2010).

[KOC00] S. W. Koch, T. Meier, F. Jahnke, and P. Thomas: *Microscopic theory of optical dephasing in semiconductors*, Appl. Phys. A **71**, 511–517 (2000).

[KOR10] J. Korn: *Influence of doping on quantum dot laser turn-on dynamics* (2010), Bachelorarbeit.

[KOR13] V. V. Korenev, A. V. Savelyev, A. E. Zhukov, A. V. Omelchenko, and M. V. Maximov: *Analytical approach to the multi-state lasing phenomenon in quantum dot lasers*, Appl. Phys. Lett. **102**, 112101 (2013).

[KOR13a] V. V. Korenev, A. V. Savelyev, A. E. Zhukov, A. V. Omelchenko, and M. V. Maximov: *Effect of carrier dynamics and temperature on two-state lasing in semiconductor quantum dot lasers*, Semiconductors **47**, 1397–1404 (2013).

[KOV03] A. R. Kovsh, N. A. Maleev, A. E. Zhukov, S. S. Mikhrin, A. V. Vasil'ev, A. Semenova, Y. M. Shernyakov, M. V. Maximov, D. A. Livshits, V. M. Usti-nov, N. N. Ledentsov, D. Bimberg, and Z. I. Alferov: *InAs/InGaAs/GaAs quantum dot lasers of 1.3 μm range with enhanced optical gain*, J. Crystal Growth **251**, 729–736 (2003).

[KRO63] H. Kroemer: *A proposed class of hetero-junction injection lasers*, Proc. IEEE **51**, 1782–1783 (1963).

[LEE11c] J. Lee and D. Lee: *Double-state lasing from semiconductor quantum dot laser diodes caused by slow carrier relaxation*, J. Korean Phys. Soc. **58**, 239 (2011).

[LIN10] B. Lingnau, K. Lüdge, E. Schöll, and W. W. Chow: *Many-body and nonequilibrium effects on relaxation oscillations in a quantum-dot micro-cavity laser*, Appl. Phys. Lett. **97**, 111102 (2010).

[LIN11b] B. Lingnau: *Many-body effects in quantum dot lasers with optical injection*, Master's thesis, TU Berlin (2011).

[LIN12] B. Lingnau, K. Lüdge, W. W. Chow, and E. Schöll: *Influencing modu-lation properties of quantum-dot semiconductor lasers by electron lifetime engineering*, Appl. Phys. Lett. **101**, 131107 (2012).

[LIN12a] B. Lingnau, K. Lüdge, W. W. Chow, and E. Schöll: *Many-body effects and self-contained phase dynamics in an optically injected quantum-dot laser*, in *Semiconductor Lasers and Laser Dynamics V, Brussels*, edited by K. Pana-jotov, M. Sciamanna, A. A. Valle, and R. Michalzik (SPIE, 2012), vol. 8432 of *Proceedings of SPIE*, pp. 84321J–1.

[LIN13] B. Lingnau, W. W. Chow, E. Schöll, and K. Lüdge: *Feedback and injection locking instabilities in quantum-dot lasers: a microscopically based bifurca-tion analysis*, New J. Phys. **15**, 093031 (2013).

[LIN14] B. Lingnau, W. W. Chow, and K. Lüdge: *Amplitude-phase coupling and chirp in quantum-dot lasers: influence of charge carrier scattering dynamics*, Opt. Express **22**, 4867–4879 (2014).

[LUE08] K. Lüdge, M. J. P. Bormann, E. Malić, P. Hövel, M. Kuntz, D. Bimberg, A. Knorr, and E. Schöll: *Turn-on dynamics and modulation response in semiconductor quantum dot lasers*, Phys. Rev. B **78**, 035316 (2008).

[LUE09a] K. Lüdge, E. Malić, and E. Schöll: *The role of decoupled electron and hole dynamics in the turn-on behavior of semiconductor quantum-dot lasers*, in *Proc. 29th Int. Conference on Physics of Semiconductors (ICPS-29), Rio de Janeiro 2008, AIP Conf. Proc. 1199*, edited by 2009), vol. 1199 of *AIP Conf. Proc.*, p. 475.

[LUE09] K. Lüdge and E. Schöll: *Quantum-dot lasers – desynchronized nonlinear dynamics of electrons and holes*, IEEE J. Quantum Electron. **45**, 1396–1403 (2009).

[LUE10a] K. Lüdge, R. Aust, G. Fiol, M. Stubenrauch, D. Arsenijević, D. Bimberg, and E. Schöll: *Large signal response of semiconductor quantum-dot lasers*, IEEE J. Quantum Electron. **46**, 1755 –1762 (2010).

[LUE10] K. Lüdge and E. Schöll: *Nonlinear dynamics of doped semiconductor quantum dot lasers*, Eur. Phys. J. D **58**, 167–174 (2010).

[LUE11] K. Lüdge, E. Schöll, E. A. Viktorov, and T. Erneux: *Analytic approach to modulation properties of quantum dot lasers*, J. Appl. Phys. **109**, 103112 (2011).

[LUE11a] K. Lüdge: *Modeling Quantum Dot based Laser Devices*, in *Nonlinear Laser Dynamics - From Quantum Dots to Cryptography*, edited by K. Lüdge (WILEY-VCH Weinheim, Weinheim, 2012), chapter 1, pp. 3–34.

[LUE12] K. Lüdge and E. Schöll: *Temperature dependent two-state lasing in quantum dot lasers*, in *Laser Dynamics and Nonlinear Photonics, Fifth Rio De La Plata Workshop 6-9 Dec. 2011*, edited by (IEEE Publishing Services, New York, 2012), IEEE Conf. Proc., pp. 1–6.

[MAI60] T. H. Maiman: *Stimulated optical radiation in ruby*, Nature **187**, 493 (1960).

[MAJ11] N. Majer, S. Dommers-Völkel, J. Gomis-Bresco, U. Woggon, K. Lüdge, and E. Schöll: *Impact of carrier-carrier scattering and carrier heating on pulse train dynamics of quantum dot semiconductor optical amplifiers*, Appl. Phys. Lett. **99**, 131102 (2011).

[MAJ12] N. Majer: *Nonlinear Gain Dynamics of Quantum Dot Semiconductor Optical Amplifiers*, Ph.D. thesis (2012).

[MAL07] E. Malić, M. J. P. Bormann, P. Hövel, M. Kuntz, D. Bimberg, A. Knorr, and E. Schöll: *Coulomb damped relaxation oscillations in semiconductor quantum dot lasers*, IEEE J. Sel. Top. Quantum Electron. **13**, 1242–1248 (2007).

[MAR03c] A. Markus, J. X. Chen, O. Gauthier-Lafaye, J. G. Provost, C. Paranthoen, and A. Fiore: *Impact of intraband relaxation on the performance of a quantum-dot laser*, IEEE J. Sel. Top. Quantum Electron. **9**, 1308 (2003).

[MAR03a] A. Markus, J. X. Chen, C. Paranthoen, A. Fiore, C. Platz, and O. Gauthier-Lafaye: *Simultaneous two-state lasing in quantum-dot lasers*, Appl. Phys. Lett. **82**, 1818 (2003).

[MAX13] M. V. Maximov, Y. M. Shernyakov, F. I. Zubov, A. E. Zhukov, N. Y. Gordeev, V. V. Korenev, A. V. Savelyev, and D. A. Livshits: *The influence of p-doping on two-state lasing in InAs/InGaAs quantum dot lasers*, Semicond. Sci. Technol. **28**, 105016 (2013).

[MEU09] C. Meuer, J. Kim, M. Lämmlin, S. Liebich, G. Eisenstein, R. Bonk, T. Vallaitis, J. Leuthold, A. R. Kovsh, I. L. Krestnikov, and D. Bimberg: *High-speed small-signal cross-gain modulation in quantum-dot semiconductor optical amplifiers at 1.3 μm*, IEEE J. Sel. Top. Quantum Electron. **15**, 749 (2009).

[MEY91] P. Meystre and M. Sargent: *Elements of Quantum Optics* (Springer Verlag, 2nd edition, 1991).

[MIK05] S. S. Mikhrin, A. R. Kovsh, I. L. Krestnikov, A. V. Kozhukhov, D. A. Livshits, N. N. Ledentsov, Y. M. Shernyakov, I. I. Novikov, M. V. Maximov, V. M. Ustinov, and Z. I. Alferov: *High power temperature-insensitive 1.3 μm InAs/InGaAs/GaAs quantum dot lasers*, Semicond. Sci. Technol., 340–342 (2005).

[MAR03] B. E. Martínez-Zérega, A. N. Pisarchik, and L. S. Tsimring: *Using periodic modulation to control coexisting attractors induced by delayed feedback*, Phys. Lett. A **318**, 102–111 (2003).

[OTT14] C. Otto: *Dynamics of Quantum Dot Lasers – Effects of Optical Feedback and External Optical Injection*, Springer Theses (Springer, Heidelberg, 2014).

[PAU12] J. Pausch, C. Otto, E. Tylaite, N. Majer, E. Schöll, and K. Lüdge: *Optically injected quantum dot lasers - impact of nonlinear carrier lifetimes on frequency locking dynamics*, New J. Phys. **14**, 053018 (2012).

[PLA05] C. Platz, C. Paranthoen, P. Caroff, N. Bertru, C. Labbe, J. Even, O. Dehaese, H. Folliot, A. Le Corre, S. Loualiche, G. Moreau, J. C. Simon, and A. Ramdane: *Comparison of InAs quantum dot lasers emitting at 1.55 μm under optical and electrical injection*, Semicond. Sci. Technol. **20**, 459–463 (2005).

[RAF11] E. U. Rafailov, M. A. Cataluna, and E. A. Avrutin: *Ultrafast Lasers Based on Quantum Dot Structures* (WILEY-VCH, Weinheim, 2011).

[ROE14] A. Röhm, B. Lingnau, and K. Lüdge: *Understanding ground-state quenching in quantum-dot lasers*, IEEE J. Quantum Electron. **51** (2015).

[SCH58] A. L. Schawlow and C. H. Townes: *Infrared and optical masers*, Phys. Rev. **112**, 1940 (1958).

[SCH88j] E. Schöll: *Dynamic theory of picosecond optical pulse shaping by gain-switched semiconductor laser amplifiers*, IEEE J. Quantum Electron. **24**, 435–442 (1988).

[SCH89] E. Schöll: *Theoretical approaches to nonlinear and chaotic dynamics of generation-recombination processes in semiconductors*, Appl. Phys. A **48**, 95 (1989).

[SCH07f] A. Schliwa, M. Winkelnkemper, and D. Bimberg: *Impact of size, shape, and composition on piezoelectric effects and electronic properties of In(Ga)As/GaAs quantum dots*, Phys. Rev. B **76**, 205324 (2007).

[SCH12e] H. Schmeckebier, C. Meuer, D. Arsenijević, G. Fiol, C. Schmidt-Langhorst, C. Schubert, G. Eisenstein, and D. Bimberg: *Wide-range wavelength conversion of 40-Gb/s NRZ-DPSK signals using a 1.3-µm quantum-dot semiconductor optical amplifier*, IEEE Photonics Technol. Lett. **24**, 1163–1165 (2012).

[SCU97] M. O. Scully: *Quantum Optics* (Cambridge University Press, 1997).

[SOK12] G. S. Sokolovskii, V. V. Dudelev, E. D. Kolykhalova, A. G. Deryagin, M. V. Maximov, A. M. Nadtochiy, V. I. Kuchinskii, S. S. Mikhrin, D. A. Livshits, E. A. Viktorov, and T. Erneux: *Nonvanishing turn-on delay in quantum dot lasers*, Appl. Phys. Lett. **100**, 081109 (2012).

[SUG05b] M. Sugawara, N. Hatori, H. Ebe, M. Ishida, Y. Arakawa, T. Akiyama, K. Otsubo, and Y. Nakata: *spectra of 1.3-µm self-assembled InAs/GaAs quantum-dot lasers: Homogeneous broadening of optical gain under current injectionmodeling room-temperature lasing*, J. Appl. Phys. **97**, 043523 (2005).

[SUG05] M. Sugawara, N. Hatori, M. Ishida, H. Ebe, Y. Arakawa, T. Akiyama, K. Otsubo, T. Yamamoto, and Y. Nakata: *Recent progress in self-assembled quantum-dot optical devices for optical telecommunication: temperature-insensitive 10 Gbs directly modulated lasers and 40 Gbs signal-regenerative amplifiers*, J. Phys. D **38**, 2126–2134 (2005).

[TON06] C. Z. Tong, S. F. Yoon, C. Y. Ngo, C. Y. Liu, and W. K. Loke: *Rate Equations for 1.3- µm Dots-Under-a-Well and Dots-in-a-Well Self-Assembled InAs-GaAs Quantum-Dot Lasers*, IEEE J. Quantum Electron. **42**, 1175–1183 (2006).

[VES07] K. Veselinov, F. Grillot, C. Cornet, J. Even, A. Bekiarski, M. Gioannini, and S. Loualiche: *Analysis of the double laser emission occurring in 1.55 µm InAs−InP (113)B quantum-dot lasers*, IEEE J. Quantum Electron. **43**, 810–816 (2007).

[VIK05] E. A. Viktorov, P. Mandel, J. Houlihan, G. Huyet, and Y. Tanguy: *Electron-hole asymmetry and two-state lasing in quantum dot lasers*, Appl. Phys. Lett. **87**, 053113 (2005).

[VIK07a] E. A. Viktorov, M. A. Cataluna, L. O'Faolain, T. F. Krauss, W. Sibbett, E. U. Rafailov, and P. Mandel: *Dynamics of a two-state quantum dot laser with saturable absorber*, Appl. Phys. Lett. **90**, 121113 (2007).

[WEG14] M. Wegert, D. Schwochert, E. Schöll, and K. Lüdge: *Integrated quantum-dot laser devices: Modulation stability with electro-optic modulator*, Opt. Quantum Electron. **46**, 1337–1344 (2014).

[ZHA10a] Z. Y. Zhang, Q. Jiang, and R. A. Hogg: *Simultaneous three-state lasing in quantum dot laser at room temperature*, Electron. Lett. **46**, 1155 (2010).

[ZHU12a] A. E. Zhukov, M. V. Maximov, Y. M. Shernyakov, D. A. Livshits, A. V. Savelyev, F. I. Zubov, and V. V. Klimenko: *Features of simultaneous ground- and excited-state lasing in quantum dot lasers*, Semiconductors **46**, 231–235 (2012).

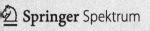